A Software Engineering Approach to LabVIEW™

Kenneth L. Ashley
- Analog Electronics with LabVIEW

Jeffrey Y. Beyon
- Hands-On Exercise Manual for LabVIEW Programming, Data Acquisition, and Analysis

Jeffrey Y. Beyon
- LabVIEW Programming, Data Acquisition, and Analysis

Mahesh L. Chugani • Abhay R. Samant • Michael Cerra
- LabVIEW Signal Processing

Jon Conway • Steve Watts
- A Software Engineering Approach to LabVIEW

Nesimi Ertugrul
- LabVIEW for Electric Circuits, Machines, Drives, and Laboratories

Rahman Jamal • Herbert Pichlik
- LabVIEW Applications and Solutions

Shahid F. Khalid
- Advanced Topics in LabWindows/CVI

Shahid F. Khalid
- LabWindows/CVI Programming for Beginners

Thomas Klinger
- Image Processing with LabVIEW and IMAQ Vision

Hall T. Martin • Meg L. Martin
- LabVIEW for Automotive, Telecommunications, Semiconductor, Biomedical, and Other Applications

Bruce Mihura
- LabVIEW for Data Acquisition

Jon B. Olansen • Eric Rosow
- Virtual Bio-Instrumentation: Biomedical, Clinical, and Healthcare Applications in LabVIEW

Barry Paton
- Sensors, Transducers, and LabVIEW

Jeffrey Travis
- LabVIEW for Everyone, second edition

A Software Engineering Approach to LabVIEW

▲ **Jon Conway**
▲ **Steve Watts**

PRENTICE HALL
Professional Technical Reference
Upper Saddle River, New Jersey 07458
www.phptr.com

Library of Congress Cataloging-in-Publication Data

Conway, Jon.
    A software engineering approach to LabVIEW / Jon Conway, Steve Watts.
        p. cm. — (Natural instruments virtual instrumentation series)
    ISBN 0-13-009365-3
1. LabVIEW. 2. Software engineering. I. Watts, Steve, II. Title.
  QA76.758.C693 2004
  005.1—dc21

                                                    2003007211

Editorial/production supervision: *Carlisle*
Cover design director: *Jerry Votta*
Cover design: *Nina Scuderi*
Art director: *Gail Cocker-Bogusz*
Manufacturing manager: *Alexis R. Heydt-Long*
Manufacturing buyer: *Maura Zaldivar*
Publisher: *Bernard Goodwin*
Editorial assistant: *Michelle Vincenti*
Marketing manager: *Dan DePasquale*
Full-service production manager: *Anne R. Garcia*

Contents

2

LabVIEW Rocks 7

3

Software Design Principles 15

4

LabVIEW Component Oriented Design (LCOD) 33

▼5

LCOD Implementation   47

▼6

LCOD Complementary Techniques   59

7

Software Engineering Essentials   105

8

It's All About Style   131

9

The Journey   141

Preface

There are many ways of designing and implementing a system. We are not trying to say that you should immediately adopt the techniques presented in this book in place of how you currently design and write software. Specifically, what we are saying is that this is how we design and implement software in real-world applications. We want you, the reader, to draw your own conclusions.

It's important to note that the authors are working engineers who pay their mortgages by writing software, not by writing books.

The Test Engineer's Perspective

Steve Watts writes—

As a normally trained test engineer I've been programming test systems for years and using many different programming languages (HPBasic, UCLA Pascal, Turbo Pascal, Visual Basic, and QuickBasic). In many of the more complex systems I have had the same experience. Doing little design up front I would plow into the coding, by the 50% stage I would normally be ahead of the game, and at the 90% stage I would be 90% complete and patting myself on the back. And then it happened!

I now use the term "the complexity explosion" small changes in the software would cause problems throughout the system. The customer would throw in "unplanned-for" changes. I could no longer picture the system clearly in my head. The last 10% of the project took another 90% of the time.

I knew something was wrong but didn't have the tools or training to explain what, why, or how. In the end I put it down to software being a pain.

A few years ago when Jon came to the company he was touting a language called LabVIEW. This became the company standard, so I had to learn it. The first application that I wrote (in a very unpleasant style I hasten to add) was a small temperature logging effort. It became clear to me that something was still wrong. True, G gave huge productivity increases over Pascal and Visual Basic, which I was using at the time, but the complexity explosion was still there, lurking in the background. I went back to Jon and discussed it with him and he introduced me to LCOD. I had never thought that there was a discipline called Software Engineering (I thought by writing software I was a software engineer), or heard of Coupling, Cohesion, or Information Hiding. OOD, OOA, and structured software design had all passed me by.

I'm the sort of person who needs to completely understand a process beyond the words, and since we were dealing with reasonably abstract concepts I struggled in the search for this comprehension. I took postgraduate courses in Software Engineering and Object Oriented Programming. I experimented with the projects I was working on, using structured software design, CASE tools, and OOA. The inherent complexity that academia applies to all things and the embracing of this complexity (out of elitism perhaps!) by the software community, led me to believe that this whole process was harder than I thought. BUT IT'S NOT!

I began to see that by applying these techniques my programs were becoming manageable, they were not increasing in complexity near the end, and I could implement late changes without reducing system robustness. Maintenance was easier and faster, customers were happy and impressed, stress levels were reduced, illness and pestilence were driven from the land, neighbor loved neighbor, and there was peace in our time.

Don't get me wrong, none of this will make a complex problem any less complex, but at least by applying these techniques you won't be making it more complex.

As software engineers we should be striving for the following:

- Deliver what we say we are going to deliver
- Deliver it when we say we are going to deliver it

- Ensure that it operates predictably
- Ensure that changes and bug-fixes do not harm the stability of the program or break the bank to implement

We should be in the business of managing complexity.

$$\text{Clever Software} = \text{BAD}$$
$$\text{Simple Software} = \text{GOOD}$$

One of our customers wrote the following testimonial (and we didn't even pay him!): "[LCOD has] made a complex test system simple, flexible, and futureproof."

Using the analogy of a journey (as we do throughout the book), we feel we have taken enough steps forward to enable us to turn around and put up a few signposts. Hopefully, these signposts will help you in your journey.

I have never regretted adding flexibility to my software, but I have always rued the times I have omitted flexibility.

The techniques presented in this book are reasonably simple to understand. We feel that someone can only successfully apply something if they understand it. Our aim is to introduce and explore the concepts of software design using LabVIEW, and to do this in an understandable and applicable manner. A lot of techniques and methodologies get bogged down with computer science and forget about the design aspects; our intentions are to always concentrate on design and hopefully translate some of the computer science.

Acknowledgments

We'd both like to thank Bernard Goodwin (Prentice Hall) and Tiffany Kuehn (Carlisle Publishers Services) for all their help. Also to our reviewers, especially John Compton-Smith for their useful and motivating comments.

Thanks to all my family and friends for being just that. Thanks to the ladies in my life, Yvette, Jasmine and Poppy without whom I would have finished this book a lot quicker. Their interruptions lighten my day. I would also like to thank the Tuesday night club and patient staff in the Red House (Whitchurch, Hampshire, UK) for the beer, tequila, food and discussions. Finally I'd like to thank anyone who has paid me money to play with these hi-tech toys.

—Steve Watts

About Prentice Hall Professional Technical Reference

With origins reaching back to the industry's first computer science publishing program in the 1960s, and formally launched as its own imprint in 1986, Prentice Hall Professional Technical Reference (PH PTR) has developed into the leading provider of technical books in the world today. Our editors now publish over 200 books annually, authored by leaders in the fields of computing, engineering, and business.

Our roots are firmly planted in the soil that gave rise to the technical revolution. Our bookshelf contains many of the industry's computing and engineering classics: Kernighan and Ritchie's *C Programming Language*, Nemeth's *UNIX System Adminstration Handbook*, Horstmann's *Core Java*, and Johnson's *High-Speed Digital Design*.

PH PTR acknowledges its auspicious beginnings while it looks to the future for inspiration. We continue to evolve and break new ground in publishing by providing today's professionals with tomorrow's solutions.

Introduction

1

Welcome to our book. We hope that the ideas and examples presented within this text give you something to work with, or at least get you to reconsider the way you currently use Laboratory Virtual Instrument Engineering Workbench (LabVIEW). Using the techniques we present in this book will make your life easier and make complex projects more attainable. We want this book to be pragmatic; we're practical people who happen to make our living using these techniques.

In every engineering discipline a good design is the biggest contributor to success; conversely, a poor design is a major reason for failure. If there is one thing we would like you to remember after having read this book, it is this fact.

What we present in the following pages is the culmination of our journey into software so far, with an emphasis on LabVIEW. There are no "Silver bullets" in this book, just the design strategies and implementation techniques that have proved to be successful for us.

Good luck in your journey.

1.1   LabVIEW Sucks

J. Conway writes, "LabVIEW sucks!" Yep, I was convinced that this stupid "picture" language I was being told to use was absolute rubbish. Tom, a colleague of mine, and I were working on a large Automatic Test Equipment (ATE) project. It was 1993, and I believe it was more luck than judgment that had led the management to choose LabVIEW over the other languages that were available at the time. My background prior to this was all text based, certainly not PC based. I absolutely hated this language, as well as Windows 3.1.

I have also encountered many other software engineers who hate LabVIEW (hate is a very strong word, but it certainly seems to sum up their feelings). So, why did I have such a hard time, as well as many other engineers? I think the answer is simple, it is very hard to program extremely badly in LabVIEW because you are forced to use the data flow paradigm. In general this means that all users of LabVIEW are being forced to program in a certain way. At least this means that all software written follows a certain pattern. To many programmers this becomes infuriating, but I think they, and earlier I, have missed the point.

Don't get me wrong, LabVIEW does not make a complete novice write superbly engineered code, quite the opposite (and hence one of the underlying reasons for this book). What LabVIEW does do is lay the foundation for how the software is constructed. For example, when looking at the world of C++, probably one of the most popular languages for the software engineer, there are many ways code can be written. If you are a C++ user you may be gasping with horror at this statement, but I feel it is true. Many people think of C++ as Object Oriented (OO), assuming that all code written is OO. This is not the case. It is quite possible to write software in C++ that bears little resemblance to any OO at all, because the engineer is not forced to use OO techniques.

What makes LabVIEW different is that you do not have an option to write software using anything but the data flow principle. In fact, if you tried really really hard you probably could get around this, but it would be too infuriating to take very far. This removes the flexibility to deviate from or invent your own system. Some may say that this stifles creativity; I say that they are talking nonsense. Any software is merely a tool for expressing a design, as a building expresses the design of an architect. What is important is how strong the software is, which can be compared to the cohesive components that make a building stay up. Imagine the total chaos if builders were allowed to use whatever materials they liked, bonded together in a manner that was left up to the individual. I think you probably get the point. When you look at the many software projects that fail, overrun, or do not meet the requirements, you can begin to see why it is such an advantage to work with underlying rules.

There are many other advantages to using LabVIEW, which will be discussed in later chapters, but before going any further it is important to grasp the concept of how a tool is used. Another big problem in the world of software is how the language is portrayed. LabVIEW has often been marketed as a software package that makes writing software easy.

Software is not easy. LabVIEW does indeed make easy things easy to do, but as with all modern languages once you go beyond the simplest of tasks the ease of software construction diminishes as a prerequisite for success. If you are only going to write the simplest of applications, then you probably do not need this book. However, if you are going to be involved in medium- to large-scale projects, the principles and techniques presented here should give you a fighting chance for success. This book presents nothing new, which is another point of immense importance. This book attempts to show you how to apply already well-defined software engineering principles to LabVIEW.

This is not a learn to program LabVIEW in two seconds book, or a bible, or an unleashed or advanced version; it is simply for engineers to be able to construct LabVIEW code with basic software engineering principles. This book is also not for someone who does not "buy into" the most important and underused aspect of software engineering, design. I do not intend to come up with a new way of doing things, just the right way. What is laid out in the following chapters is what I have put into practice in the projects that I have been involved in using LabVIEW. There is a big "aha" factor for me for a lot of what I present, and I feel that most of what I am attempting to show here will be of immense help.

The underlying purpose of this book is not to show tricks and tips on certain abstracted problems, but ways of constructing LabVIEW code from a proper design. If you do not like the sound of this, then put the book down and walk away because there is a lot more to come. As the title says, this book is intended as a guide to present items that I have personally used, chewed over, and hopefully come up with a way of tackling the issue. However, I am not arrogant enough (some may disagree) to suggest that this is the only way, or in every case the best way, of doing things. I hope that the book will be an ongoing piece of work that builds a community of members who are interested in the advancement of software engineering within the LabVIEW arena. My intention is that this book will set in motion the continual improvement of how LabVIEW software is constructed by inviting as much constructive criticism as possible.

1.2   Don't Buy This Book

Sometimes you buy a book that makes wild claims about why you should buy the book, and what it is going to do for you. We would like to invert this and give reasons why you should not buy this book, if any of the following is true:

- You have never read a book about software engineering.
- You're not receptive to new design ideas.
- You don't own any books about software engineering.
- You do not like working in a team, when necessary. You hide your code from prying eyes and claim it as your own property.
- You claim to have eight years' experience writing G, and no one can teach you a damn thing (in fact, you probably have two years' experience repeated four times).
- You think that all the other books written about LabVIEW are better than this one (joke).

This is all a bit tongue in cheek, but our point is that if you haven't been nodding your head in agreement with what we've stated so far, then this book is probably not for you.

1.3   The Soap Box

What is wrong with the way LabVIEW code is written? Why is there so little material on how to use proper software engineering techniques with LabVIEW? The answers, we think, are in the way LabVIEW is portrayed. We have often complained to all those who would listen that the marketing of LabVIEW is a huge millstone. In our opinion, LabVIEW is an extremely feature-rich, mature, and robust software development environment. Unfortunately, LabVIEW is portrayed, and regarded, as an easy tool for those wishing to write applications but who have little or no software engineering or programming experience.

It is true that easy things are easy to do in LabVIEW, as indeed they are in many other comparable languages such as Visual Basic. But, go beyond a quick data acquisition (DAQ) application and often chaos ensues. Why? Because a large majority of the LabVIEW community have little or no experience in software engineering or programming. The result is amateurish applications at best. Why anybody thinks that someone with no experience can create an application in LabVIEW is beyond us. To us it's akin to giving someone with absolutely no building experience all the latest power tools and asking them to build a house. Surely it's easy with all those tools, isn't it? No, it's not!

So, here it comes, guess what? Writing software is not easy. Another metaphor we like to use is tiling a bathroom wall. If all you have to do is place one single tile on the wall, the job seems relatively simple. Okay, you still have to get the tile on the wall straight and put enough adhesive on the wall for it to stay up, but all in all quite simple. However, if you were to put up five tiles things can get a little trickier. A slight error has a compound effect on the other four tiles; you have to start considering the gaps between the tiles; you have to also ensure that every tile is square with its neighbors; and the wall may not be even so you will have to pad out some tiles with adhesive. Still it is only five tiles, so it won't be that noticeable. However, if the wall consists of 500 tiles and the first tile is one degree off center, then the 500th tile will be miles off. It should also be clear from this metaphor that the bigger the job, the more considerations come into play.

What this illustrates is why small jobs are easier than large jobs, because any errors made at the start are amplified if not spotted straight away. For instance, an ATE is going to be based around an oscilloscope connected via a General Purpose Interface Bus (GPIB). Rather than use one common interface, all the commands to the oscilloscope are scattered throughout the four corners of the application, and anytime the oscilloscope needs to be communicated with a bit of G is slapped in. Now, we get to the end of development and a fundamental flaw in the frequency of a measurement comes to light. This frequency has to change from a range of 100 Hz to 1,000 Hz to a new range of 10 Hz to 2,000 Hz. If the scope had been in a single component with a common set of interfaces then the change would be simple. However, because the software in this case wasn't written with this in mind, all the code that talks to the scope now has to be hunted down and modified. In a small application this may only be a few instances; in a large application this could be thousands.

This example is very simplistic, but it actually happened on a project one of the authors (Jon) was involved in. For months after the modification it was found that some measurements were being taken over the wrong frequency span because another call to the scope had been missed. Think what it would have been like if another change had to be made, and then another. If the whole ATE had been written with little attention to a good design, then each attempted change would manifest itself into more errors—chaos. Sound familiar? To a greater or lesser extent this problem is common in a lot of projects, and is a fundamental cause why systems can be flaky and hard to maintain.

So, what is the cure? The strange thing about this question is the simplicity of the answer. A complete understanding of the problem domain, a good design, and well-constructed software will ensure a robust, easy to update, and maintainable application. All things considered, it will also enable the application to be delivered on time and hopefully exceed the expectations of the user or customer.

1.4 What This Book Is

With thanks to John Compton-Smith for the clarification.

In both of the two online LabVIEW communities, there are regular requests from members who are seeking software engineering texts so that they can improve their own work.

Typical LabVIEW users are not trained in software engineering techniques. They are not likely to pick up a book on software engineering or software quality unless forced to do so. We want to address this group by emphasizing a simple methodology that brings the main points of software engineering into play. The use of the LabVIEW Component Oriented Design (LCOD) method is an important building block to understanding the software engineering principles that will be discussed throughout the book.

This book doesn't have the depth usually found in software engineering books. This is because we aim to cover the points made in the level of detail that a practical software user might require and no more. What we have hopefully achieved is an approach to the subject that is direct and nonacademic.

There are a number of heavy academic books on software design and software engineering. Most of them will improve the way you produce software and most will have little appeal to engineers who just want to get on with solving their next problem. Similarly, we haven't padded the book with entry-level chapters for beginners since the documentation shipped with LabVIEW is top quality and it is pointless to replicate it.

Another important point to consider is that although the case study concentrates on Test Systems, the methodology can be applied to any software written in LabVIEW.

1.5 Companion Web Site

Go to the companion Web site located at http://authors.phptr.com/watts/ for full source code and book updates.

LabVIEW Rocks

2

2.1 Why Does LabVIEW Rock?

We're going to be controversial here. LabVIEW is the best general-purpose language on the market today.

It's not sold as that, it's sold as a test and measurement tool for nonprogrammers. Even seasoned professionals hedge their bets by saying that it is the best tool for certain jobs. In our experience, for a large proportion of projects that come to us, LabVIEW is the best language to program in. The arguments against this statement are reasonably weak. One such argument is that you cannot easily control the mouse pointer image in LabVIEW. Big deal, that is hardly a project stopper; we've yet to find a customer who actually cares about some of the Windows niceties and frills. National Instruments (NI) is making LabVIEW more available through real-time initiative, and we say there will come a day when LabVIEW will break out of its specialist market and go mainstream.

Why LabVIEW is so good is a contentious issue. We spent some time researching various studies and came to the conclusion that we should make up our own study. We're obviously keen on LabVIEW, so is this a keenness based on the peculiar loyalty programmers have with a particular language? Our interest is purely in productivity. We've tried many languages and methodologies and none come close to LabVIEW in pure productivity. So what has LabVIEW got that these languages haven't, and what do we mean by productivity? Let us discuss productivity first. The following is a list of important characteristics of a software language. If the language scores highly in all of them it is reasonable to assume that the language is going to be productive.

- Readability

  How much of the problem domain is visible on a single computer screen? How much do you need to learn before you can understand the source code? How complex are the primitives of the language? (Primitives are the basic building blocks, if the basic structures and data types are too numerous or not clearly defined the language will be difficult to use.) How clear is the structure of the program? Does the language support information hiding?

- Writability

  How quickly can you transfer your designs into the real world? Can you separate the design of the User Interface from the underlying source code? Are small changes in the program made by correspondingly small changes in the code?

  Can you break the source code into chunks to hide complexity? Are the interfaces to these chunks clearly defined? Can a hierarchy of these chunks be built up, pushing complexity down to lower levels?

- Editability

  It's much easier and faster to edit something than to create something new. A blank sheet of paper is the worst thing to start with when designing. What does the language offer in the form of templates, and how easy is it to modify existing code?

- Reusability

  How much do you have to write from scratch? Can you build on the development environment so that your own code can be reused as seamlessly as the language's own built-in functionality?

- Understandability

  What does the language offer when trying to visualize the problem? Is the customer likely to be involved in the process, and will the customer be able to understand the general process?

So how does LabVIEW shape up, and is the list we used the complete story?

- Readability

  LabVIEW is hierarchical in structure so there is no reason why the complete top-level Virtual Instrument (VI) shouldn't be visible on a single screen. The basic primitives and structures are quite small in number (the graphical nature helps to navigate, and the context help is invaluable when learning). The debugging tools are great for understanding the operation of the program.

  The basic structures (Sequence, Case, While Loop, and For Loop) ringfence their code in a way that couldn't be clearer. This allows the structure of your code to be clearly defined. If you've trawled through tons of nonindented Pascal, C, or BASIC trying to find the end of a loop you'll appreciate this feature.

  The data flow paradigm is simple to understand. Someone with a reasonable amount of experience can decipher even the most unpleasant program. One criticism leveled at LabVIEW is that the data flow paradigm constrains it in some way. These constraints help readability. OO languages such as Visual Basic (VB) become too much of a personal interpretation of the problem domain. Their lack of constraint means that there are 1,001 different ways to solve a problem. For these languages the solution to a problem is more a case of looking it up and cutting and pasting something than working it out.

- Writability

  When talking to LabVIEW developers about why they like LabVIEW, one thing that is quite often mentioned is that programming in LabVIEW is fun. This would seem to indicate that it scores highly in the writability stakes. If it were difficult to convert your ideas to reality using LabVIEW, you would find it to be less fun. If you had people on your back about your performance, you would also view LabVIEW as less than fun. So do we derive this fun from dragging and dropping icons onto a block diagram? We hope not. We think the fun comes from the graphic design aspects of the language. You create a pleasant User Interface and that makes you feel good. You then create the source code that operates the User Interface and that looks nice too. It's all very satisfying.

  The fact that you can design the User Interface separately from your source code is something to sing about. It now has this in common with quite a few languages (VB, Delphi, and National Instruments' C-based environment (CVI) plus there are quite often display builder packages available for languages like SmallTalk and Java). Can functionality be added to the displays of these other languages with the customer sitting behind you? Usually not. This is an enormous advantage; the fact that the customer is involved and not excluded is an excellent start. The capability to produce prototypes rapidly and, therefore, inexpensively to help gather requirements is one of the unique advantages of LabVIEW.

Are small changes in requirements echoed in small changes in the source code? If you design your code right in LabVIEW we suggest that even large changes could be accommodated. This changes the whole dynamic of programming. If you are not running in fear of change, you can plan later, in less detail, and get away with it.

The development environment offers productivity help in other ways, for example, LabVIEW will not allow you to run if there are obvious errors in your code. You can define the inputs of any subVIs so that LabVIEW insists you connect it correctly. This saves time during development, removing the compile>>run>>debug>>recompile methods of old. This only helps with syntactical errors and not logical errors, but any help is good.

LabVIEW is a hierarchical language and this allows complexity to be managed by breaking the program into chunks. The interface and architecture of these chunks, or subVIs, can be clearly viewed in the hierarchy display.

- Editability

  One of the espoused advantages of OO is that you can add functionality without affecting the existing functionality (in OO this is achieved by using inheritance). Inheritance is a double-edged sword, and the price of inheritance is readability. The program becomes such a personal interpretation of the problem that it can be quite difficult to pick up someone else's work. That's before you even start to think about interface inheritance and multiple inheritance.

  LabVIEW gives reasonable capabilities where templates are concerned, with Control and VI templates available. Code fragments can be reused by the use of Merge VIs. There is also a vast set of example software available with the standard package.

- Reusability

  One reasonably critical argument of why LabVIEW was so productive stated that it was because of the huge array of reusable code available with the development environment. This is an advantage, but shouldn't all software languages have the same advantage? Visual Basic, for example, must have at least 50 times more reusable code available.

So why is the reusable code that comes with LabVIEW so usable, and why does it leverage so much advantage? Clarity of interface is one thing to consider. You will be more likely to reuse the available code if it is easy to use. All VIs are defined by their inputs and outputs. These inputs and outputs are easily viewed by popping up on the context help. "Popping up" is LabVIEW jargon for right clicking and selecting from the menu.

Your own and other people's VIs can be loaded into the development environment and used as seamlessly as the language's built-in functionality.

There are data types that can be defined that improve the capability to reuse further and these will be discussed later in the book.

- Understandability

Programming is problem solving and you solve a problem by visualizing it. Good programmers can visualize the many states and interactions of their program in their heads and end up having an enlarged node on their brain. This enlarged node grows at the expense of the surrounding areas, these being the areas responsible for memory, small talk, and fashion sense.

Another conclusion of the research is that LabVIEW, being graphical, aids in this visualization. Here's an analogy.

When getting directions in your car the person will give you a list of landmarks, turnings, and place names. If this list is more than five items long you will more than likely forget something and end up lost. After a couple of times you will be able to get to your destination without referring to the list of directions. You will be using the visual part of the brain, which has converted the list of words to the actual objects.

In programming if you are thinking in pictures rather than words, you will be bypassing a level of abstraction in your head. Programming is hard enough on your brain without having to translate text to pictures.

But this is not the complete story. The fact that you are doing something graphical makes the whole programming event something that is watchable. This opens up the whole process, allowing interested parties to get involved from the start. This input is invaluable for developing to your customer's need, and the customer's input is invaluable for understanding the problem.

S. Watts writes a final story. I was moaning to an NI representative about faultfinding a program thrown together by a nonprogrammer fresh from the LabVIEW Basics course. This program was truly horrible and was being used to control a piece of hardware that I was not familiar with on a really old PC. I fixed it after about an hour.

The conversation went something like this:

Me: What on earth are you turning out from these LabVIEW courses? I've just had to fix the most horrible program ever.

NI Rep: Oh yeah, what was the problem?

Me: Oh, it was badly written, using a piece of kit I was unfamiliar with on a really old rubbish PC. It took me nearly an hour to fix it.

NI Rep: An hour, so what was the problem?

Me, losing ground fast: Well, it was really ugly and I struggled.

NI Rep: Yeah, but you fixed it, right?

Me: er . . . Yeah.

NI Rep: LabVIEW's great isn't it!

And he was completely right. If it had been any other language, there is absolutely no way I could have fixed it in an hour.

On a footnote, the NI LabVIEW Basics courses are valuable because they run through the capabilities of the language, but they will not teach you how to design software.

2.2   What Advantages Does This Bring to the Developer?

Forget the fact that you are a test engineer, scientist, or production engineer; you are being asked to develop software. Go on, call yourself a software engineer. How much time do you spend developing software per week? The additional productivity that LabVIEW brings is a bonus, but it's not the whole story.

Before we talk about everything else let's linger on productivity a bit longer. This is more than just building software quicker and cheaper, it's about opening up new areas of business, making bespoke software available where it was previously too expensive. Developing quicker allows you to react to changing requirements quicker. This gives the advantage of being at least in step with your customer and, in some cases, even ahead. There aren't many languages where you can say that!

The story about the ugly program discussed earlier highlights another advantage of LabVIEW: maintainability. In traditional languages maintenance can be the most expensive phase of a project. LabVIEW projects generally involve a fair bit of maintenance, in the test environment new tests are specified, new hardware added, and operating systems get upgraded (!!). The fact that it isn't a huge problem in LabVIEW means that it is reasonably transparent, but it's a big advantage.

LabVIEW gives you the tools to write very flexible code. Issues that used to stop a project in its tracks can be quite effortless in LabVIEW. You may not have used them all or thought about writing with flexibility in mind, but they are there and waiting to be used. You will never regret putting flexibility into your software.

Finally, reusability. Every reasonably decent software engineer has a personal library of code that he or she can call upon. LabVIEW allows the programmer to simply incorporate your own code into the development environment.

2.3 How Can Good Design Leverage These Advantages?

We feel that the real power of LabVIEW has yet to be fully expressed because of design weaknesses, not language weaknesses. The techniques described here will help you to take productivity, maintainability, and reusability to the next level.

Generally, you are not paid for your prowess and in-depth knowledge of a language and its tools, but rather for solving your customer's problem. This book will demonstrate how good design can make your life simpler and your software better by using LabVIEW as it was designed and not forcing it to do something it isn't designed for, or introducing half-understood techniques from the advanced palette. While the techniques described here will bring instant advantages to you by simplifying your source code, the real benefits will be felt a year or two after you first start applying the techniques. This will be when you've tamed the complexity beast and you stand atop your software like a programming colossus.

Software Design Principles

3

Before looking at the technical aspects of what makes a good software application we should establish what we think constitutes good software.

A good software application should:

- Enable small changes in users' needs to be accommodated by small changes in the code
- Be delivered on time and on budget
- Function mostly to expectations
- Be easy to use
- Be maintainable
- Perform well
- Fail gracefully
- Be secure
- Work reliably

3.1　Why is Software Complex?

. . . software seems like malleable stuff, most programs are actually intricate plexuses of brittle logic through which data of only the right kind may pass.[1]

—W. Waytt Gibbs, "Software's Chronic Crisis"

Software can be thought of as a machine that has many states, with these states being linked to or triggered by other states. Consider that even the most basic application consisting of a user interface, a bit of hardware, and maybe some data that needs filing can have more possible states than we would care to list. This is why it is possible, even inevitable, to be overrun by our software. Added to this inherent complexity is the perception that creating software is easy, and changing someone else's even easier.

It has been proven that software maintenance is a significant proportion of the total cost of the software. Most programmers hate picking up other people's software more than being asked to do paperwork. The reasons for this are relatively easy to define.

Other peoples software is often:

- Overly complex.
- Badly tested.
- Written in a peculiar style.
- Poorly documented.
- Too clever.

On top of that are the reasons we dare not discuss:

- Changing other people's software is always more problematic than we anticipated
- It always takes too long
- We always end up owning the entire application afterward
- A small change can ripple through the application causing death and destruction in its wake

[1]From "Software's Chronic Crisis," by W. Wyatt Gibbs, *Scientific American*, September 1994.

The last point is interesting and is related to how the many states of the application interact. If our software has many states that are dependent on other states, small changes will propagate themselves throughout the application. This interaction of states is called coupling, a fundamental part of software design, which will be discussed more thoroughly later.

As software engineers we are in the business of managing complexity. So let's all agree on the following:

- Simplicity is good . . . complexity is bad.
- Good design simplifies complex problems.
- Decomposing large complex systems into manageable chunks is an effective method of complexity management. The interfaces to these chunks should be understandable enough to allow them to be used without referring to their implementation.

What weapons can we employ in our battle against complexity?

- Coupling
- Cohesion
- Information hiding
- Abstraction

The following analogy serves two purposes: it introduces you to the ACME Widget Company and illustrates coupling and cohesion in an example that the majority of the readers will be familiar with.

There are four products made at the ACME widget factory, the ubiquitous Widget, its successor the SuperWidget, the top of the line MegaWidget 2000, and the all new method of measuring widgets, the Widgetometer.

Regarding the factory layout (as seen in Figure 3.1), it can be seen that a product's journey through the shop floor is complex, and the distance traveled by the subassemblies and parts is long. Managing a factory such as this one is complicated because any product, part, or subassembly could be anywhere in the factory.

You can think of the factory as a software application, parts going into stores would be input data and shipped products would be output data. The production lines would be top-level components.

Before we improve matters let's define cohesion and coupling and then apply them to the factory layout.

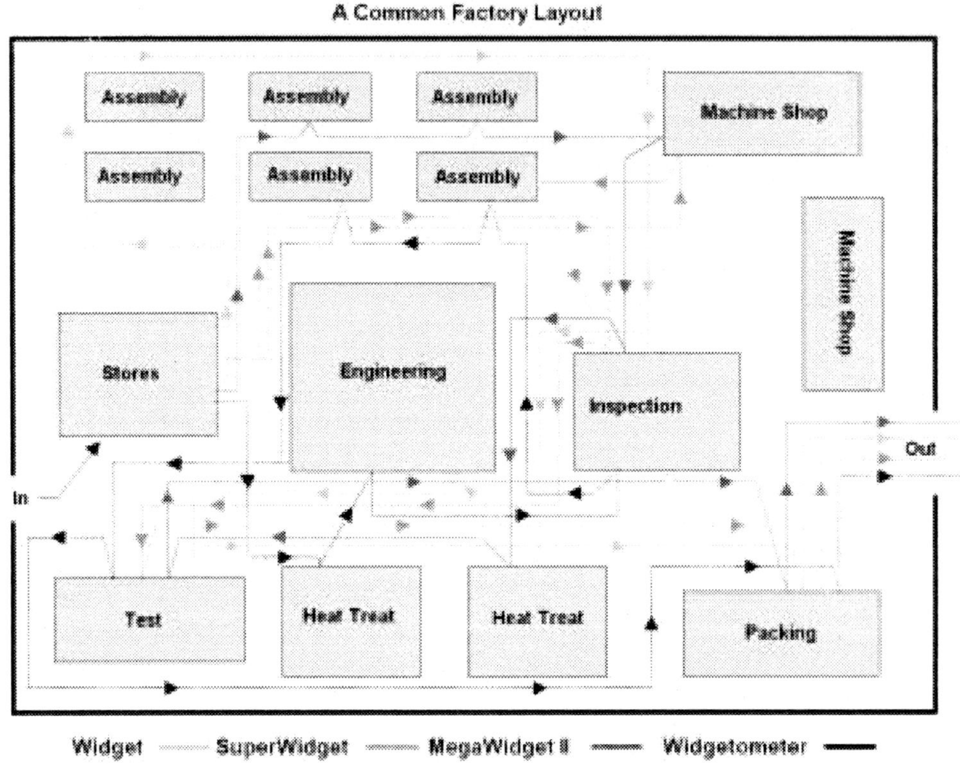

Figure 3.1
Disorganized factory.

3.2   Coupling and Cohesion

Coupling and cohesion are terms used to describe the effective modularity of a component. They are central to designing robust, maintainable, and reusable software.

Dictionary Definitions

> **Cohesion—1** the act or condition of sticking together; **2** a tendency to cohere; **3** *Physics* the sticking together of molecules of the same substance
>
> **Coupling—1** That which connects

Software Academia Definitions

Cohesion—A measure of how strongly the elements within a module are related. The stronger the better.

The idea of cohesion in software design first appeared in the mid-1960s, courtesy of Larry Constantine. The term was acquired from sociology, where it applies to the relatedness of humans within groups. For our purposes we can think of cohesion as collecting all similar functionality in one component.

Coupling—A measure of the degree of independence between modules. When there is little interaction between two modules, the modules are described as loosely coupled. When there is a high degree of interaction, the modules are described as tightly coupled.

There are various types of coupling and cohesion defined, but this is outside the scope of this book. In a well-designed component we are looking for loose coupling and strong cohesion.

Now that that's cleared up, back to the factory. Consider that the top-level processes for the factory are deliver a Widget, deliver a SuperWidget, deliver a MegaWidget 2000, and finally deliver a Widgetometer. Each subprocess is dependent on various other subprocesses, and each production line is intertwined with all the others. This is tight coupling. A symptom of tight coupling is the movement of parts and subassemblies around the shop floor, which is closely analogous to the data in software. Each department in the factory is not single-mindedly producing for one production line. It would be very difficult to establish what work is being done where. This is weak cohesion.

It can be seen in Figure 3.2, that by reorganizing the factory into flow lines for each product type and storing parts on the line at the point where they are to be used, the following improvements are made:

Movement of products, parts, subassemblies, and people around the factory has been reduced.

Management of the production lines has been simplified.

The flow line principle turns each production line into a minifactory, independent and unaffected by any other areas of the shop floor. A production line could be removed, modified, or added without affecting any of the others. Production lines therefore exhibit very strong cohesion and are loosely coupled to one another. As described earlier, the factory is analogous to a software application and the production lines are individual components. The internal

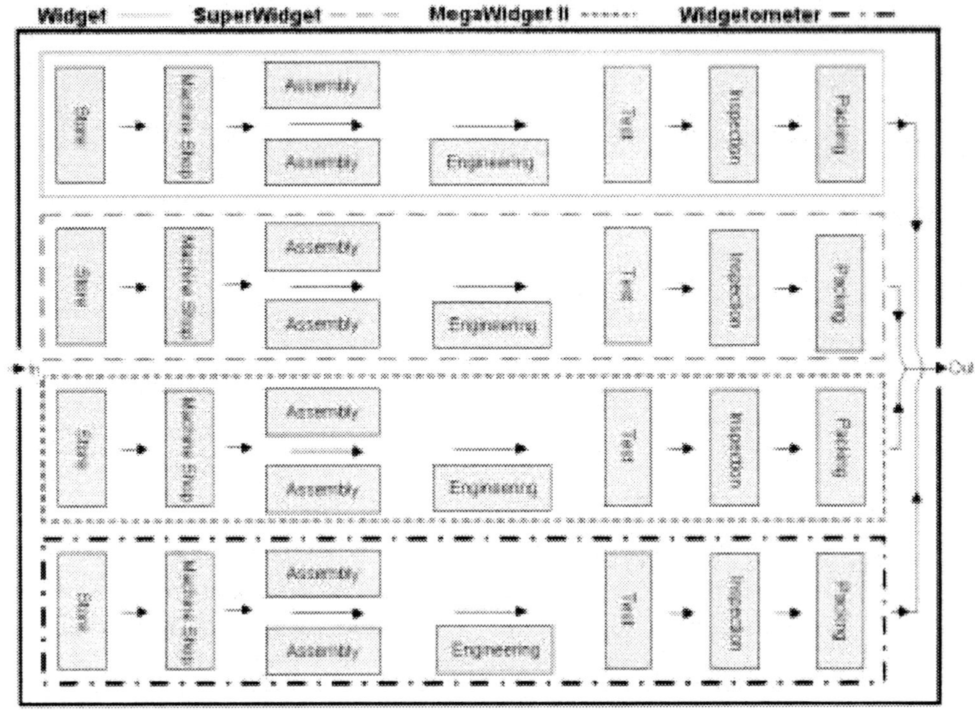

Figure 3.2
Reorganized factory.

production line processes could be thought of as private subroutines and the parts as data.

Now consider the following. When you accelerate your car, you are not thinking about how the gasoline is pulled from the tank into the cylinders where it ignites, pushing the piston down, which turns the crankshaft faster, which turns the gears faster, which then turns the driveshaft, which transmits power to the wheels. When you steer, you don't think about the power steering, or for that matter braking and changing gears. Cars are very complex, but as far as the user is concerned this complexity is hidden away behind a simple interface. Not only is it a simple interface but it's a fairly common interface, and luckily automobile manufacturers have agreed on its format and stuck to it. We could otherwise find ourselves in the following situation.

The diagrams in Figure 3.3 illustrate four possible ways of controlling a vehicle; they are four interfaces to the services provided by the vehicle.

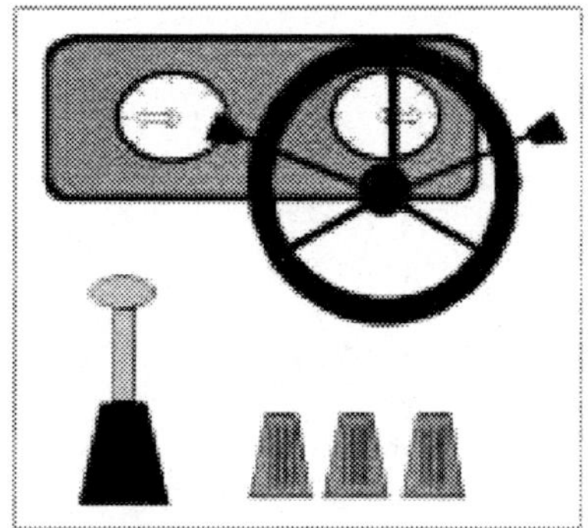

It'll never work

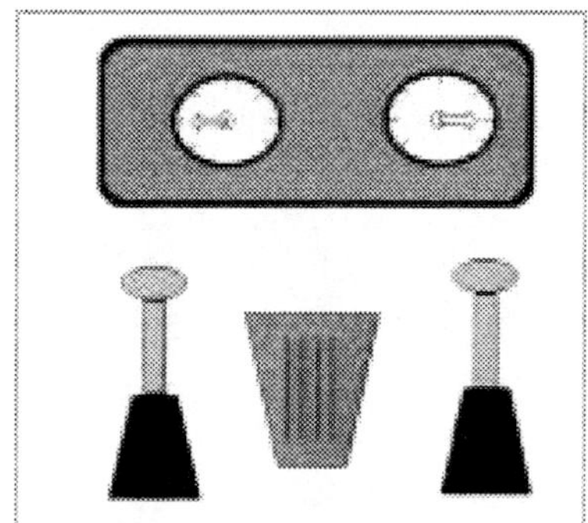

It's a tank

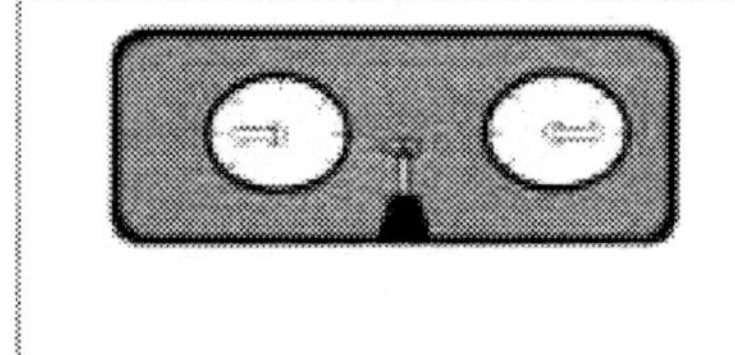

The future perhaps

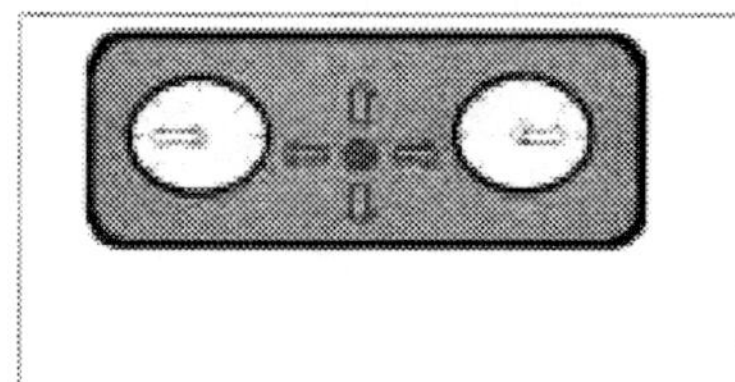

Car for the gameboy
generation

Figure 3.3
Possible car
interfaces.

Transferring from one mode to another would involve learning a new combination of controls. This would not only be tiresome but could be dangerous. The likelihood of an accident is greatly increased during the learning phase.

3.3   Information Hiding and Encapsulation

Dictionary Definition

> **Encapsulation**—To enclose in a capsule

Software Academia Definitions

> **Information Hiding**—Information hiding is used to hide complexity behind a simple interface.

> **Encapsulation**—The process of compartmentalizing the elements of an abstraction that constitute its structure and behavior; encapsulation

serves to separate the contractual interface of an abstraction and its implementation.[2]

The information hiding principle was first proposed by D. L. Parnas in 1972 as a criterion for decomposing a software system into modules. The principle states that each module should hide a design decision that is considered likely to change. Each changeable decision is called the secret of the module.

Information hiding allows an implementation to be hidden behind an interface that doesn't change even if the implementation does. The concept of encapsulation is related to information hiding. The data and the operations that manipulate the data are all combined in one place. They are encapsulated within a module.

Information hiding is often touted as a major reason why OO software is more robust. This is true, but with a little effort we can also benefit from implementing information hiding in our system. In a nutshell, all information is hidden away, the only way to get to the data is via its component interface.

So now that we understand the main concepts for software design, how do we implement them in LabVIEW?

3.4   Examples of Coupling, Cohesion, and Information Hiding

Something to keep in mind when looking at the examples and when considering your own designs is that coupling, cohesion, and information hiding are related. By improving information hiding you will generally be improving the coupling and cohesion.

The following examples should help in clarifying the concepts when applied to LabVIEW.

3.4.1   Bad (Tight) Coupling

The Front Panel in Figure 3.4 has many different controls and indicators, and their attributes and values will all need changing. The test and logging software behind this front panel was large and complex (1,000+ VIs). The

---

[2]From Brooch, *Object-Oriented Analysis and Design with Applications,* © 1994 Benjamin Cummings Publishing Company Inc. Reprinted by permission of Pearson Education, Inc.

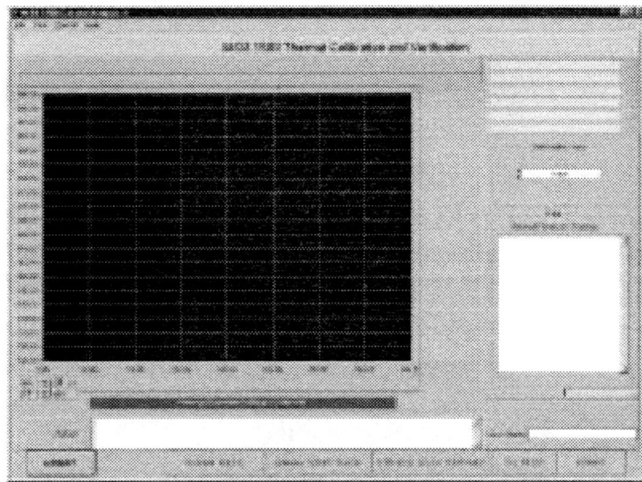

Figure 3.4
Coupling front panel.

Figure 3.5
Spaghetti code.

graphing was user-definable and it also had to accommodate all the customers' off-line analyses. The wiring diagram could have ended up like that shown in Figure 3.5.

The term *spaghetti code* was first used before LabVIEW was developed. One advantage of graphical programming is that it becomes obvious if your program is heading down this route.

In LabVIEW 6i National Instruments tackled this problem by introducing the control reference. This is essentially a means of helping with the real estate issues caused by tightly coupled Front Panel controls and indicators.

Figure 3.6 demonstrates another good indicator of tight coupling, the amount of wires going in and out of a VI. If a component is strongly cohesive and exhibits good information hiding, it will generally not need more than four or five inputs and a similar amount of outputs.

A question worth asking is, Is the data we're passing essential to the problem domain? There are numerous examples of references and pointers being passed around the software for no better reason than to tie the VIs together.

3.4.2 Good (Loose) Coupling

Going back to the example of the Front Panel as shown in Figure 3.4, another fly in the ointment on this particular project was that the customer wanted the data handling and User Interface done by us off-site (actually we wanted to be off-site). We suggested decoupling the display using message sending through a queue. We implemented a display controller VI that was continuously polled by the Main Display VI. Display updates and attribute changes were sent to the Display Controller and queued in local memory.

This gave us the following benefits:

- It allowed us to design, test, and implement the User Interface off-site.

- It simplified testing since each state of the display was defined and known.

- It allowed a change of Display State to be triggered from almost anywhere within the software.

- It allowed changes to be easily and robustly implemented.

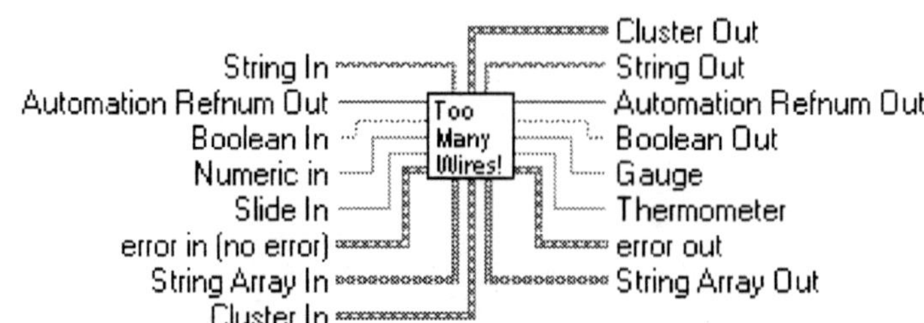

Figure 3.6
An indicator of
tight coupling.

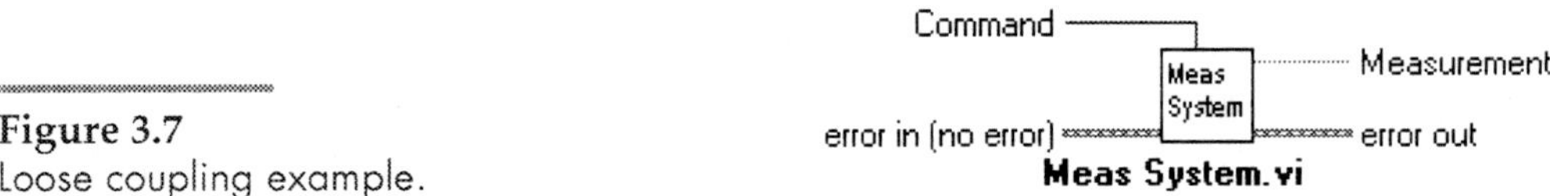

Figure 3.7
Loose coupling example.

A fuller discussion of these User Interface techniques can be found in Chapter 6 on complimentary techniques.

Another example of loose coupling could be a measurement system component. This component conducts all of the measurement functions in a test system. Its interface could be as simple as the example shown in Figure 3.7. Stripped down all you want to do is tell the component what measurement you want and for it to output a number and its status.

3.4.3 Bad (Weak) Cohesion

The classic example of weak cohesion is if you take a diagram and ring-fence it with your Create SubVI tool, but you do it in a completely haphazard way. If you show no regard at all for the operations that you are ring-fencing, you will have a system that is coincidentally cohesive. From a maintenance point of view a system partitioned like this would be a complete nightmare!

Think about the test system that we discussed earlier. It consists of an oscilloscope and a signal generator. It needs to conduct 30 individual tests. We create 30 VIs (named Test1.vi to Test30.vi). Each one has the GPIB calls to the scope and sig.gen. inside them. A reasonable test of your design is to consider the implications of change. For example a change of test frequency could involve changing specific calls in each one of the test VIs. Missing one call could invalidate the test.

You can improve cohesion simply by collecting like functionality. In this case you could make an Oscilloscope VI, Sig Gen VI, and a Test System VI. The Oscilloscope and Sig Gen VIs encapsulate all the functionality of their respective pieces of equipment. The Test System VI encapsulates all of the states of the test system. Essentially, it would become a wrapper for Oscilloscope and Sig Gen VIs.

3.4.4 Good (Strong) Cohesion

The VI shown in Figures 3.8 and 3.9 contains all the functionality required to simply control Word 97's basic functions using ActiveX. Figure 3.10 demonstrates how it is used.

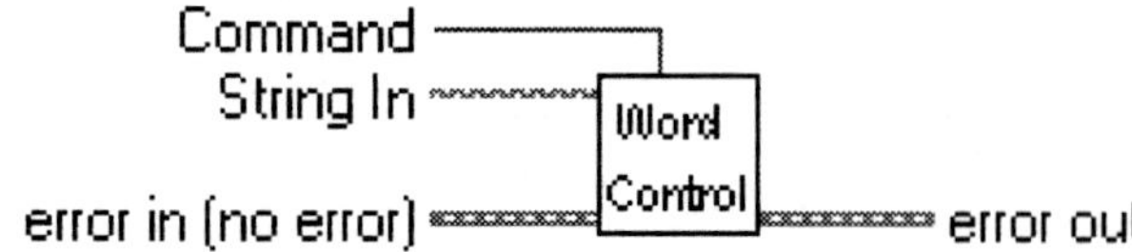

Figure 3.8
Word Control connector.

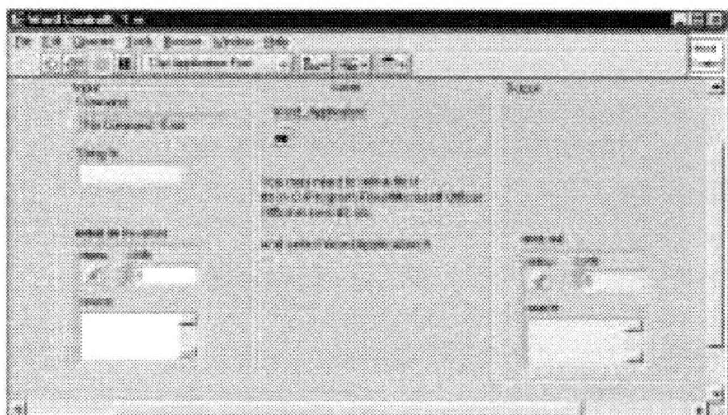

Figure 3.9
Word Control Front Panel.

This is an example of strong cohesion because the component only deals with Word 97 and completes all of its functions single-mindedly. If you have any problems with Word 97 you know where to look. If you want to test the component, its functionality is all in one place, and the tests can be easily defined. Adding functionality is purely a matter of adding a command and a new case, none of the existing functionality is changed.

3.4.5 Bad Information Hiding

In this example we have a system that uses a Digital Input and Output (DIO) card to control lights and relays. Usually DIO cards work by ports and bytes, so here is one implementation that is used to switch a relay on and a light off (to add to the confusion let's put in a bit of negative logic for the light [true = off, false = on]). The relay will switch power to a unit under test, and the light will tell the user that the unit has power.

We'll be using the NI DIO-XX card that hasn't been newly developed.

Figure 3.11 doesn't look too bad, maybe a little bit of confusion with the negative logic. Now let's multiply it 100-fold. The system is much bigger, there are a lot of switches, actuators, and lights, and the person who wrote the above innocent-looking software has left. They've changed some of the driver

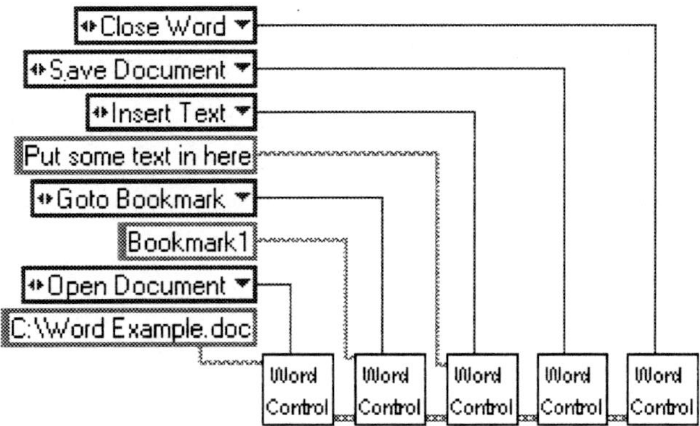

Figure 3.10
Word Control usage.

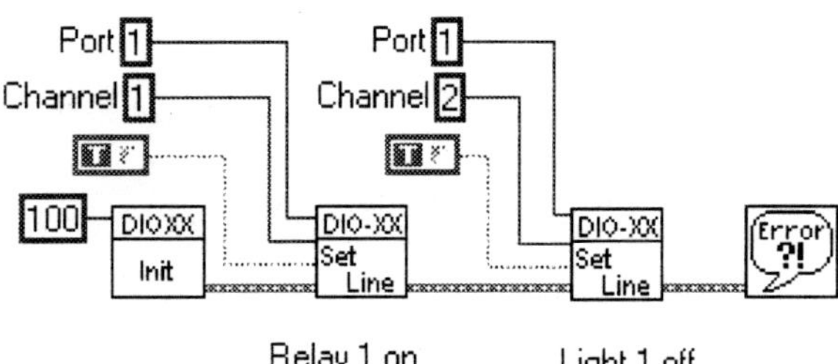

Figure 3.11
Bad information hiding.

hardware and some of the logic of the drivers. You've now got to go wading through the code changing channels and line settings to suit. If your predecessor has been somewhat lax in his or her comments, it could be a bit of a job.

DIO drivers shouldn't be complex, they are only 1s and 0s. If all the software has been written like this, there may be some real complexity in the system.

3.4.6 Good Information Hiding

When thinking about information hiding in components, we need to think about the job that the component is doing and not the way it is going to do it. So revisiting the previous example, let's think about what the component will do.

- Switch Power to unit X on.
- Switch unit X Power indicator on.

- Switch Power to unit X off.
- Switch unit X Power indicator off.

The initialize function doesn't have anything to do with the problem domain, the component should decide for itself whether it needs initializing.

Let's redesign the DIO-XX component as shown in Figures 3.12 and 3.13.

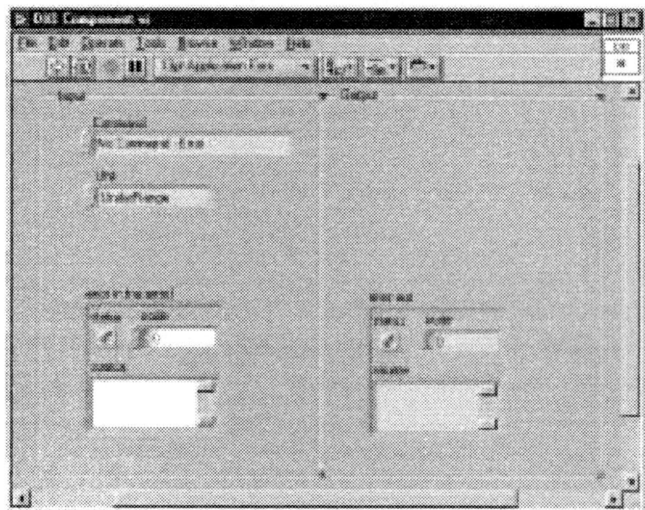

Figure 3.12
DIO Component Panel.

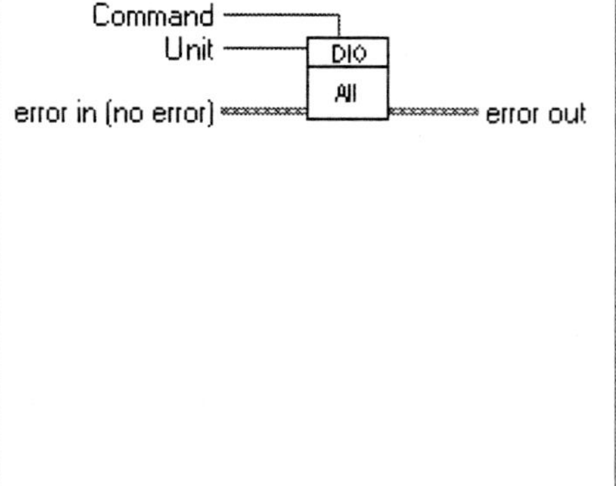

COMMANDS	
No Command Error: Should throw an error loudly in the real world	

COMMANDS
No Command Error: Should throw an error loudly in the real world
Switch Units Power On: Actuates Relays for selected unit on
Switch Unit Power Off: Actuates Relays for selected unit off
Switch Unit Power Indicator On: Turns selected unit's light on
Switch Unit Power Indicator Off: Turns selected unit's light off

Figure 3.13
DIO Component commands.

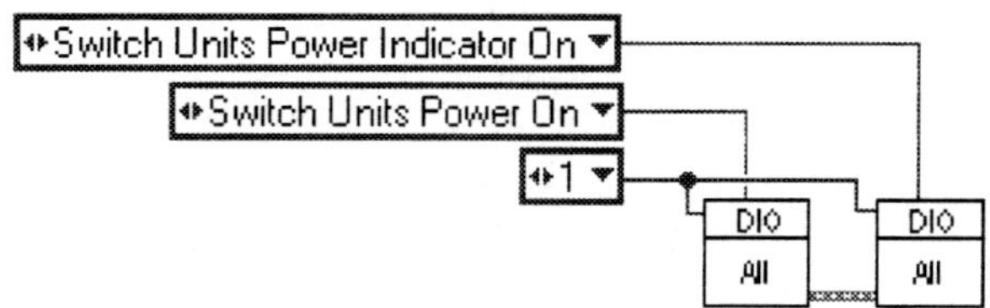

Figure 3.14
Good information hiding.

Don't worry too much about how this will be implemented for now, all will be revealed later. The implementation is less important than understanding that the component has been redesigned to be more in terms of the system than in terms of the hardware. We are not worried about ports and channels because these are implementation details. What we are interested in is the problem domain (i.e., what actually needs to happen). The redesigned diagram to complete the functions of switching a unit and its indicator on will now look like Figure 3.14.

3.5 Abstraction

Dictionary Definition

> **Abstraction**—A mental concept used to simplify a complex problem

You will come across this term a lot with regard to software design, so it is worth exploring it further.

When you save some text to a file, you will think of it in the terms of sending a string to the "Save to File.vi," instead of thinking about placing the 1s and 0s onto spare sectors of your hard drive. You are using abstraction to visualize the problem at a manageable level of complexity. The "Save to File.vi" method is at a far more abstract level than the 1s and 0s method.

In software design there are two types of abstraction.

> **Functional Abstraction:** LabVIEW provides the facility to break up your design into subVIs. This hierarchical structure provides functional abstraction. A top-level function can consist of many subfunctions, and these subfunctions can have subfunctions of their own. This is the main way of breaking down the complexities in a problem.

> **Data Abstraction:** Computers deal in 1s and 0s. Programming languages provide a level of abstraction beyond that by giving us more elaborate

data types like Integers, Real Numbers, and Strings. LabVIEW also gives you facilities to define your own data types using clusters. Clusters allow you to group related items together, simplifying the moving of data around your application.

By encapsulating and hiding data within our VI we can take abstraction to similar levels as that provided by Object Oriented Programming (OOP). Using message sending as the only way to access and change the data allows us to implement abstract data types. Abstract data types are the basis for good modular design. The following example should help illustrate the advantage of correctly using abstraction in your designs.

We want a component that controls the switching for a measurement system. The underlying technology is a hardware card that turns relays on or off. A low level of abstraction would be the subVIs used to initialize and set or clear the ports on the card. These subVIs could be used as is, and every operation in the measurement system could have the relevant subVI being called to set and clear the required relays. This is shown in Figure 3.15.

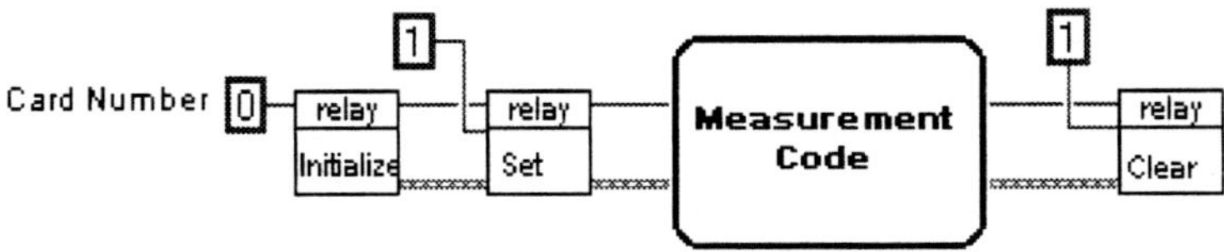

Figure 3.15
Low-level abstraction.

A higher level of abstraction is to wrap the relay VI in a component that abstracts the details of relays and channel numbers. We are now talking in terms of the connection made from a system perspective. The Switch VI is used to wrap the relay VIs. All of the relay VIs are now encapsulated in one VI. There should be no way to access the relay VIs from outside the Switch interface. The usage of these new VIs is shown in Figure 3.16.

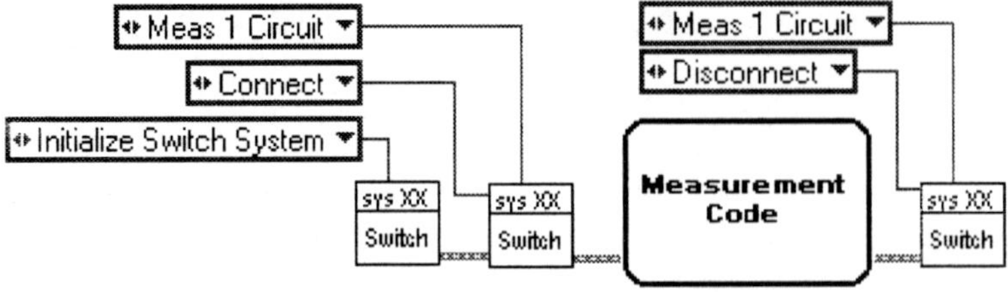

Figure 3.16
Higher level abstraction.

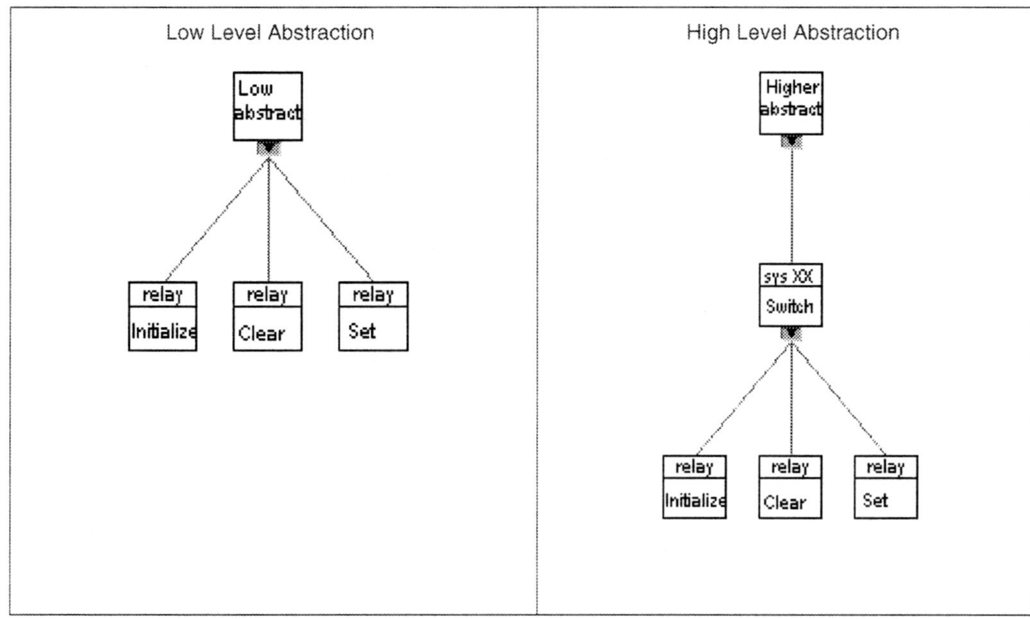

Figure 3.17
Abstraction hierarchy.

So how does all this benefit our design? Well, let's compare the hierarchy diagrams in Figure 3.17.

The higher level of abstraction will allow you to make changes to the switching system without actually changing the Switch VI interface. You could swap for a different relay card, for example, or change the switching configuration or the relays called. But the underlying structure of the software will remain the same. The extra layer of abstraction protects the software design from changes. Also, the higher level of abstraction makes your program easier to read. There is no ambiguity about what you want to happen to the switches when you send a command. Ambiguity in software is bad; ambiguity breeds bugs!

LabVIEW Component Oriented Design (LCOD)

4

Why LCOD? LCOD is just an acronym. It provides a convenient label that encompasses what we are trying to get across. So let's start with a statement. In our opinion Component Oriented Design (COD) is the best approach for developing software in LabVIEW.

COD is applicable to all software design, and certainly the same problems present themselves when beginning the design process—irrespective of the chosen language that will be used to express the design. The problem we have to solve is to take the requirements for a system and create the application. This book is about using LabVIEW to express the design, and that design being a component oriented approach, hence LCOD. So why use components at all?

Often, at project inception, there is no clear definition of the boundaries that exist in the pieces that will make up the application. So, developers naturally will pick on the interesting bits. What happens is that the bits that get picked become the driving force for the rest of the application. What transpires is not elegant.

If scant attention is paid to all components that could make up a system, clarity in the solution is diminished, if not lost. Have you ever looked at a program in which it is not obvious where one function ends and another begins? All the software is intertwined, and as it grows the relationships become more complex and the application more brittle. Eventually, even minor changes cause catastrophic failures. Identification of offending code or problems is no trivial task, and most importantly where to look is not obvious, meaning the whole application may have to be scanned to locate the problem.

Our task is to identify the cooperating components that will make up the system, then make them as autonomous as possible. The strength in the application is built on the loose coupling that we define between our components. This point is probably hard to grasp if you haven't used it before, and often with simple ideas it is hard to appreciate the benefit until used.

At this stage you may well think, "Well, that's okay for small projects, but we work on large projects and a couple of components aren't going to help us!" This is true, a couple of components aren't going to solve the whole problem, but they will solve a small part of it. What you will get at the highest level are three maybe four components. Then, each component is further decomposed into components within components, and so on.

From what has been stated previously, we know that components that are cohesive and loosely coupled are our best bet for a robust application. Well that's all fair enough, but how the heck do we go about identifying the components? Well, a little OOD, a little top-down, a little bottom-up, and a little experience.

First let us look closer at the components themselves.

4.1 Components

The idea of components is not new to LabVIEW or many other languages. There are three parts to LCOD: cohesion, coupling, and information hiding. All of these will be used in constructing components. These terms are as old as the ark as far as software engineering is concerned, but there is precious little discussion of them in the LabVIEW arena.

Some of you will notice similarities with Object Oriented Analysis (OOA). We feel we should clarify the point that LabVIEW is not an OO language and we're not trying to make it into one. The techniques described here predate OO and are methods for good modular design. When applied to LabVIEW these techniques improve an already powerful paradigm, data flow.

The idea of software components is far from new, they were first proposed in the NATO conference on software engineering in 1968 by Doug McIlroy. He realized that using libraries of commonly required utilities gave many of the benefits associated with reuse. These early software engineers were inspired by other industries that used components, and they put forward the proposal that an application could be built from software Integrated Circuits (ICs). Software components are similar to electronic components that are wired together to form larger components, with wires corresponding to the connection of data in software. This corresponds nicely with the wiring metaphor in LabVIEW. These points of interconnection between components can be thought of as the interface to the component. The interface is the only way of doing anything with the component. This interface should be explicit enough to ensure that no assumptions could be made about its implementation.

This gives us a set of requirements for what constitutes a component.

- A component must provide a clear specification of what services it is prepared to offer.
- The only way a component can interact with other components and the rest of the application is through its declared interfaces. It must encapsulate all of its data and processes behind this interface.
- The component should be independent enough to be tested in isolation.
- The components and software that use a component must rely only on their defined interfaces and specified operations.

To summarize, a good component does its duties clearly and concisely, has a simple interface, and hides its complexity from the outside world.

4.1.1 So What Is a Component?

It proved very difficult to provide a definitive answer to this, so we'll make up our own. In our terms a component is made up of VIs that provide a service or services through a simple interface. A component will exhibit the following features:

Strong cohesion—This is provided by the collection of VIs.

Information hiding—The functions of the subVIs are hidden from the outside world.

Loose coupling—This is through the simplistic interface provided for the rest of the system to realize the services.

Figure 4.1 illustrates a typical component interface. Using messages to command the component actions and optionally taking inputs and delivering

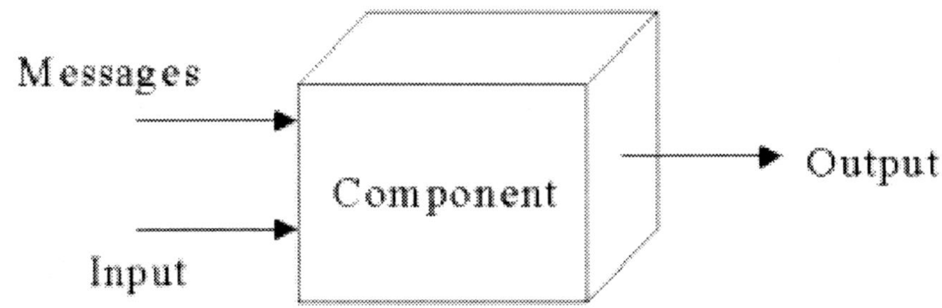

Figure 4.1
Component.

outputs. Why are components better than what you are doing now? For a small application the payoff from LCOD is small (perceived). However, for medium-sized projects and larger the rewards are enormous, if not critical to the success of the project.

A common current method of ignoring the design and concentrating on churning out a working system usually ends up with large diagrams, lack of structure, and spaghetti code, all of which lead to a complexity explosion and unpleasant amounts of stress. Side effects to this are a lack of reuse and a lack of analysis of what has been done.

Also, think how many times that you've written and rewritten routines. Add to that everyone else in the world doing the same thing. Wouldn't it be a more productive world to select and use an already written and tested component?

By talking about components we are forcing the software engineer to consider coupling, cohesion, and information hiding (even though he or she may not know it). What we are also doing is enabling a large task to be broken down into meaningful "chunks." This aids the design process, simplifies estimation, and is the first step toward breaking a complex task into more manageable parts. The abstraction of the problem into the top-level components gives us the starting point from which the rest of the project will proceed.

4.2   Design

The design of your system has important implications throughout the life of the project. LabVIEW is a software language like any other, the rules and lessons learned by other software practitioners are just as applicable in LabVIEW as they are in C, Pascal, ADA, SmallTalk, Java, or any other language. The graphical nature of LabVIEW gives your source code high visibility, making bad design harder to hide!

Your source code is a reflection of your programming ability and a public document of your professionalism. Your most critical audience will be the engineer who picks up your project after you.

The costs of poor design are:

- Missed deadlines
- High maintenance costs
- Poor reliability
- Unpredictable application behavior
- Stressed customers
- Bankruptcy, broken homes, misery, and pestilence

4.2.1 Object Oriented Design (OOD)

If you have not heard of Object Oriented Design (and associated languages), then it is probably best you put this book down and start getting out more, or maybe consider an alternative career. An OO approach to software has become a very religious experience. In some quarters unless a language or technique is claimed to be OO, it is dismissed. Let's take a closer look.

Classic OO Lesson

Object Orientation is good because you can model real-world objects using it. These principles are so easy to understand you probably understood them when still in the womb. Even a small piece of fruit could understand them. The emperor's new outfit looks fantastic!

The picture in Figure 4.2 explains it all, now go forth and write great software. But, you could also have dogs and cats as:

    class: pet
    class: animalsThatSitInYourGarden
    class: thingsThatMakeMeSneeze
    subclass: *Canis* and *Felis*

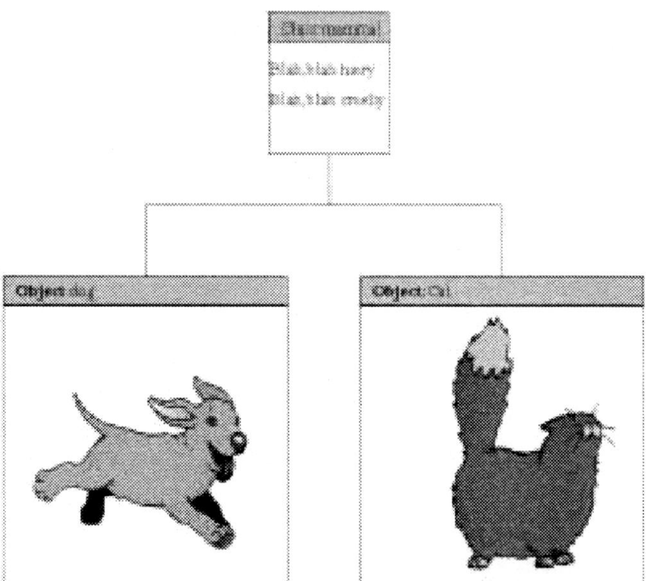

Figure 4.2
Typical OO explanation.

Although the picture and description accurately model the complexities of my dog, she is exceptionally stupid, and they do little to illustrate the real-world complexities of designing software. Modeling complex systems using OOD is still complex, we want our money back!

Flippancy aside, OO Design, Analysis, and Programming have brought a lot to the table of software design. The characteristics of OO encourage modularity, encapsulation, information hiding, loose coupling, and strong cohesion. These are all the things we want for our components. However, a badly designed OO program is as difficult to maintain as a badly designed procedural-based program.

So what do we do with OO in LCOD? We use the parts of OO that help us the most. To define OO concepts would be a whole other book, so we will try and keep this brief and simple. As you should already know the pivotal step in OO design is the identification of classes. Classes are defined by the identification of real-life and abstract objects, the operations and methods that the class will react to, and the data it will operate on. Typically, you never modify the data that belongs to a class, you only request that the class carries out the predefined operations on it. This is exactly the behavior we want our components to have. When we define a component we will hide the data implementation behind an interface; thus anyone calling our component can only interact with the data and services provided by the component via the defined interface (cohesion, coupling, and information hiding).

How do we identify possible components? A technique often employed for identifying classes is something called a noun-verb parse. Basically, look at the requirements and consider each noun and verb as a possible candidate for a class. That all may sound a bit like hard work, but you probably already do it anyway without realizing. An obvious example would be oscilloscope. If you see this noun in the requirements then it is a prime candidate for a class called oscilloscope, or in our case a component. We could offer loads of examples that exhibit nouns and verbs, but apart from being a bit boring it is probably better for you to go back and examine previous projects to see if you have already been doing this. Bear in mind that this is also iterative, and you probably won't get all your components in one go, especially in large projects.

Another technique that is more relevant to the projects you will most likely be working on (test, monitoring, or control systems) is to look at the physical entities that will exist in your system. Measuring devices, environmental control, and transducers are all prime candidate components (or classes).

A great benefit to us is that when we have identified our components, future change (and there will be plenty as the project progresses) is confined to the components themselves, whereas the services they provide will remain relatively static. This means that changes are insulated from the other components that use their services. On the face of it this probably does not sound like such a great feature, but it is probably the biggest payback from adoption of LCOD.

The one thing that OO does is it allows us to look at the problem at a high level of abstraction, but is not so good when coming to low-level structuring.

4.2.2 Top-Down Design

Top-down design sounds like you start at the very top of your application and start decomposing it into routines. This can be the case, but it is just as applicable when we take our components, which we have identified using OO techniques, and want to decompose them into the modules and inner components that will make them function. The main thing here is that we as mere humans can only concentrate on a certain amount of detail at any one time. Thus, decomposing in this manner makes the task more manageable for us, without having to deal with too much detail.

The basic way we use top-down design is to take whatever we want to decompose and look at it from the top, split it into components and look at those new components, then split them, and so on. When do you stop? Probably when it would become easier to write the code than carry on decomposing.

A prime example of where top-down design comes into its own is in designing menus. In more complicated applications there can be a complex menu system that allows the user to navigate around the application.

4.2.3 Bottom-Up Design

We consider this to be the opposite of top-down design, how obvious is that! Well, maybe not quite that obvious. Sometimes top-down and OO techniques can be at too high a level in abstraction, so to get started we use what we may already know. For instance, we know all the hardware that we are going to use, but we're not sure how it will be used by the rest of the system. By looking at what we have available we should have a better chance of figuring out what services we are able to provide. The key concept here is that we decompose using top-down but compose using bottom-up. Or, in top-down we start at the sharp end and bottom-up we start at the fat end (no pun intended). Figure 4.3 puts this into a graphical context.

Either way, both bottom-up and top-down are extremely useful in attacking the design problem. Sometimes you can try and tackle the problem using OO, not get it right, then try top-down, not get it right, and then try bottom-up. Essentially, they all give you the tools to attack the problem from all angles.

So, we've covered a little OOD, a little top-down, a little bottom-up, but what about a little experience?

4.2.4 Design Patterns

How many times have you had a design déjà vu—that feeling that you have solved a problem before not knowing where or how?[1]

—E. Gamma, R. Helm, R. Johnson, J. Vlissides, Design Patterns:
Elements of Reusable Object-Oriented Software

Experience is something, in our humble opinion, that cannot be argued against as being the major contributor in solving problems. Often the ability

---

[1] From Gamma, Helm, Johnson, Vlissides, *Design Patterns: Elements of Reusable Object-Oriented Software,* © 1995 by Addison-Wesley Publishing Company. Reprinted by permission of Pearson Education.

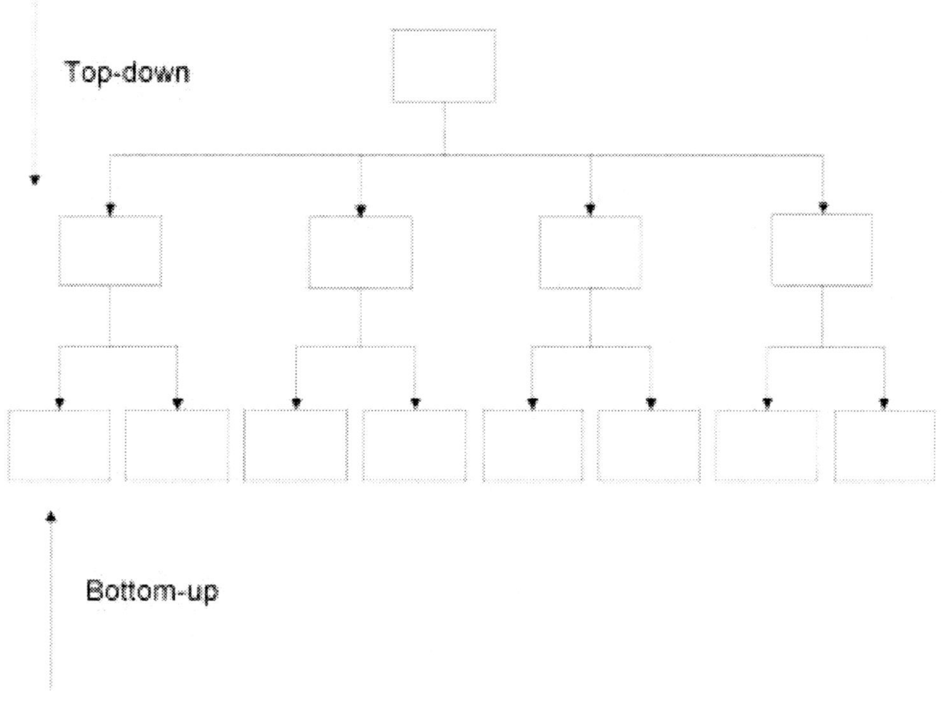

Figure 4.3
Top-down and bottom-up designs.

to solve a problem is based on our memory of previously solved problems. This means that we recognize similarities between the problem we have solved previously and the problem we find before us. This is because the problems have a pattern.

Recognizing patterns in software is something most people will do quite naturally without realizing it. However, realizing it, documenting it, and then using it has huge payback. Most discussions of patterns in software are confined to OO, which is a shame since patterns are applicable to any software and design methodology.

So what is a pattern?

Dictionary Definition

Pattern—A model or original used as an archetype. A plan, diagram, or model to be followed in making things.

Our Definition

Design Pattern—A pattern is the repeated occurrence of a solved problem.

Giving your software modules a limited interface by using message sending gives uniformity to your code that allows patterns to be observed. This observation of patterns in our code came as a direct result of using LCOD. We discovered that even working separately and from scratch on different systems there were still marked similarities in our code.

There are low-level patterns (a state machine is a design pattern), but we are going to concentrate on the higher level patterns that can be applied to complete systems.

So what do you get from the identification and use of patterns? It has been said that the difference between good and bad designers is that a good design does not solve every problem by applying first principles. They cheat!

Following is a list of what using patterns can do for you:

- Reuse of design
- Reuse of components
- Questions to ask (helps with requirements gathering)
- Cycle of reuse. The use of patterns encourages the reuse of components. (You can add more functionality to reused components because they are identified as useful and usable, and you already have the context in which they are going to be used.)

Take a test system for example, if you break it down right you will find it is very similar in structure to most other test systems. Each test system has a lot of scope for reusing code. For example handling the User Interface, Handling Data, Reporting, Error Handling, and executing the program. To illustrate the advantages this brings let's look at each item in the list above.

Reuse of Design

It is much easier to edit a design than to create a new design. Most novels are actually written to a pattern (tragic hero, triumphant underdog, etc.). Sometimes the plot pattern is more obvious than others, but it is still there.

Reuse of Components

Reusing the design simplifies and enables the reuse of components. Plugging in and adapting existing components is usually easier when there is an existing proven framework to hang them from.

Questions to Ask

This is a good one. If you have a design pattern and a set of components that have actions and attributes to define you can use this as a basis to ask very specific questions of your customer. Requirements are far easier to gather when based on specifics.

Cycle of Reuse

Applying a reusable design encourages the reuse of components, and reusing components improves the visibility of design patterns.

4.2.5 Pattern Examples

Top Level: User Interface>> System>>Hardware

User Interface: Menu Submenu Tree Pattern

Data Handling: Section Keyed Data Handling Pattern

Hardware: Control>>Drive>>Read

We'll look briefly at a couple of pattern examples next.

UI Controller>>Message Queue Pattern

One of the problems with LabVIEW is that it is difficult to decouple the display from the actual program execution. This tends to make diagrams flatter and larger than they should be. It also encourages the passing of data down the hierarchy where it isn't necessary. This results in inflexibility in design. The UI Controller>>Message Queue Pattern (Figure 4.4) draws its inspiration from the tried-and-tested techniques employed in most Graphical User Interfaces (GUIs), which is the passing of display states as messages onto a message queue. The advantages of doing it this way are as follows:

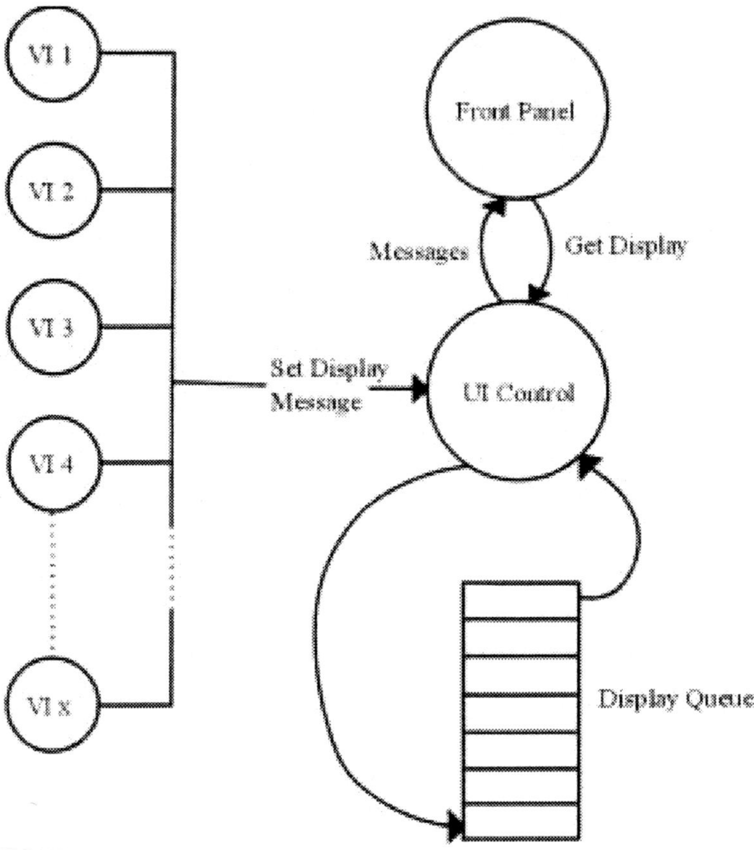

Figure 4.4
UI Controller>>Message Queue Pattern.

- The display can be changed more or less anywhere in the program hierarchy.
- The User Interface can be designed and tested isolated from the rest of the program (by using a simple test stub).
- The definition of the display is a matter of describing all the different states of the display.
- Adding new functionality does not weaken the structure of the program. It's just another message and corresponding action.

The one disadvantage is the additional memory and processing overhead required, but this would only occur on very slow computers, and there are strategies that can be used to reduce the amount of messages to economic levels.

Control>> Drive>>Read Pattern

Regarding the method, Figure 4.5 describes the Hardware actions in gradually more detail, starting with the Hardware VI, which has commands that describe the tests or actions being conducted in system facing language. A good example would be "Read Signal Response" for unit 1. This may involve switch setting, signal generation, power supply setting, and retrieval of a waveform from an oscilloscope. But as far as the Hardware VI is concerned it is a simple command. Hidden beneath the simple interface of the Hardware component will be some kind of state machine. One of the states will correspond to the request to read the signal response for the unit in question. It will then use a combination of the Control, Drive, and Read components to connect the unit and test equipment, set the power supply and generate the signal, and finally read and return the measured value.

Going to the next level then introduces the Control, Drive, and Read VIs. Sometimes you won't need all of them, so exclude any that aren't applicable, we don't mind. Again the commands you define for these components will look toward the system. So in the example mentioned previously the Control component would have commands for connecting the Unit, the PSU, Signal

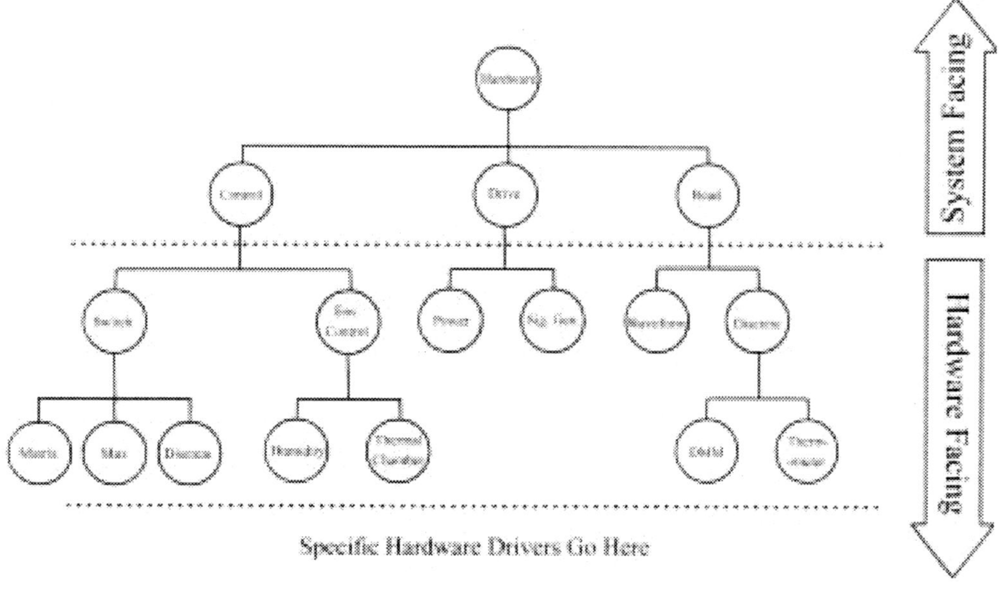

Figure 4.5
Control>>Drive>>Read Pattern.

Generator, and the selected measuring equipment to the test system. The Drive component would have commands to provide the signal and set the PSU. Finally, the Read component would have commands to return the required measurement.

> **Control:** Environmental Control, Switching
>
> **Drive:** Signal Generation, Power supply setting
>
> **Read:** Multimeter, Oscilloscope, Thermometer

A complete discussion of patterns would fill a whole other book (see the footnote at the beginning of this section). However, understanding that they exist and applying them should become second nature.

LCOD Implementation

5.1 Component Mechanisms

We should consider the inherent requirements for all components.

- All of the components, public functions, and data should be accessible via a simple interface.
- We should be able to add/delete/modify our component's actions simply.
- Any modifications should have little effect on the overall software design.
- The component stores its own state locally and persistently.
- The component should initialize itself.
- Errors are handled by the component.
- Inputs and Outputs should verify themselves.

5.2   Message Sending

Message sending is the means of controlling your component. The calling component will send a message to the called component telling it what to do.

There are various methods you could employ to send messages to your components, but we use enumerated types that are simple and self-documenting. In fact, the biggest bonus when using enumerated types is that you can drop your component on your block diagram, right-click on the input, and pop up a complete reference of the components' actions and attributes.

5.2.1   All About Enumerated Types

The LabVIEW online guide gives a comprehensive overview, however, for those too lazy here's a summary:

- An enumerated type control can be thought of as a drop-down box

- Enter additional items by using the labeling tool and pressing Shift-Enter

- The enumerated type data type is U8, U16, U32, selectable from the Representation palette. The terms U8, U16, and U32 refer to the number of elements the enumerated type can hold (U8 = 256, U16 = 65,536 and U32 = lots and lots)

- Wiring to a Case Structure makes the case display a string rather than a number

- Arithmetic operations (except Increment and Decrement) treat enumerated type as their number representation

- Increment and Decrement rotate at start and end

- A number is converted to the closest enumeration; out-of-range numbers are set to the last enumeration

5.2.2   101 Things to Do with an Enumerated Type

Well seven actually.

Format into String

Converts the selected enumeration to its text representation (in the example shown in Figure 5.1 the output string would read "Error State").

This is useful for storing states in a file or database.

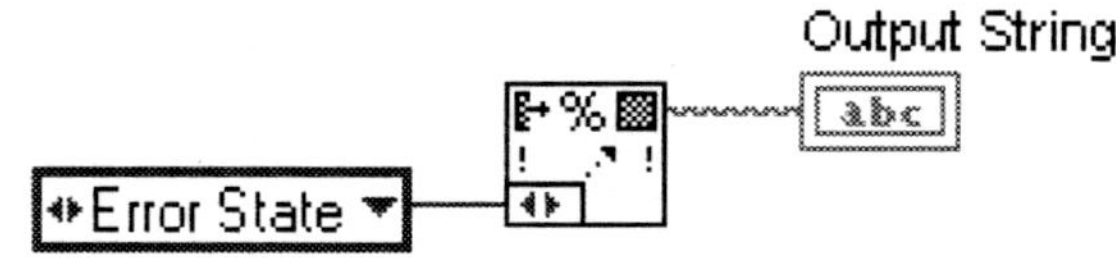

Figure 5.1
Format enumerated type into string.

Scan from String

Converts the input string to its representative output. In the example shown in Figure 5.2 the output enumeration would be set to "Test 1", if there is a corresponding list item, or "Error State" if not.

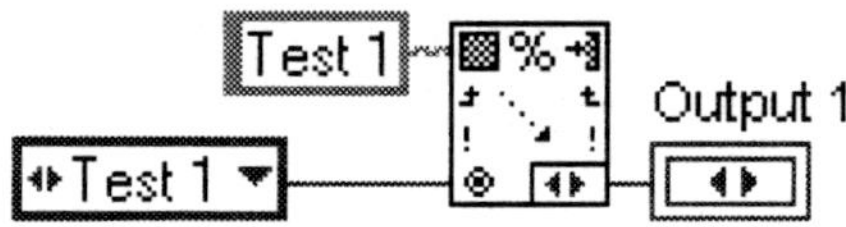

Figure 5.2
Scan enumerated type from string.

This is useful for retrieving states from a file or database.

Typecast

The example in Figure 5.3 converts the number to its enumerated type without making ugly coercion dots (Output String 2 would be set to "Test 2" in this case).

Care must be taken to ensure that the numeric is matched to the enumerations representation. If not, it may sit on "Error State".

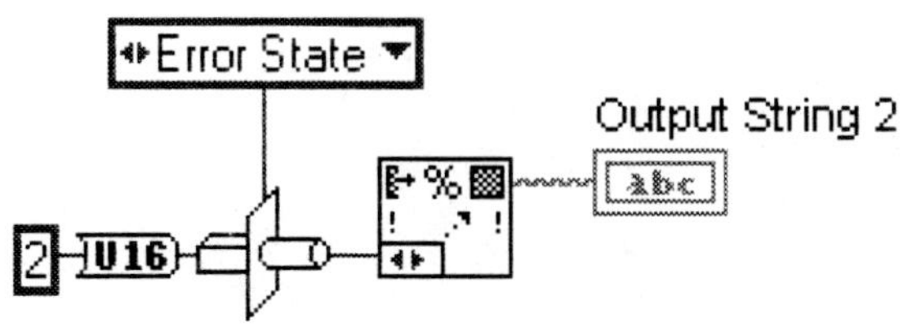

Figure 5.3
Typecasting enumerated type.

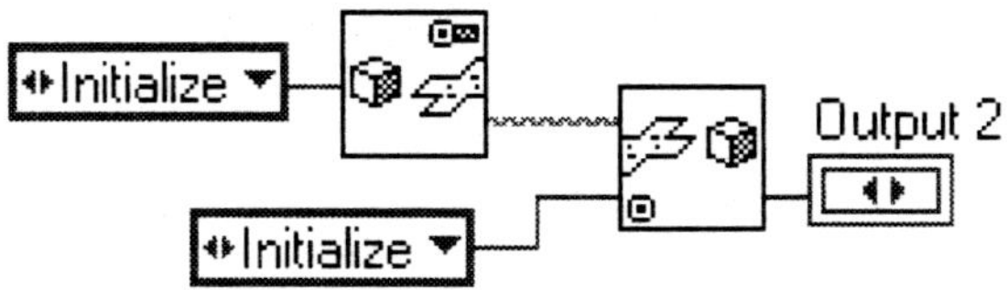

Figure 5.4
Casting to a string and casting from a string.

Casting and Uncasting

Figure 5.4 shows an enumerated type being cast to or from a string. This is very useful for sending commands through ports. For example, a real-time system could use it as a method of sending commands to and from a PC client to the real-time server using the Transmission Control Protocol/Internet Protocol (TCP/IP) functions.

Case Structure

Gives similar multiple selection functionality to a Select-Case statement in Pascal and Basic, or a Switch in C and Java. Using an enumerated type as in Figure 5.5 improves a system's readability and helps with self-documentation.

Firing a Sequence of Events

By putting the enumerated type in an array and allowing a For Loop structure to auto-index, a sequence of events can be driven. In the example shown in Figure 5.6 they go in sequence, but any order could be selected. From this simple structure a wealth of opportunities arise. For example, the array could be generated from a database or file, allowing a completely configurable system in very few steps.

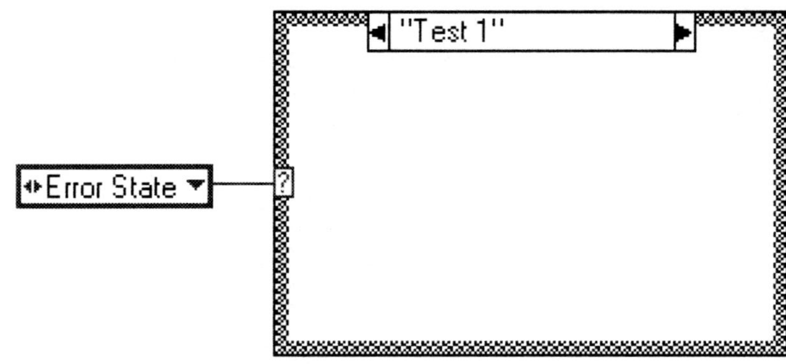

Figure 5.5
Case structure.

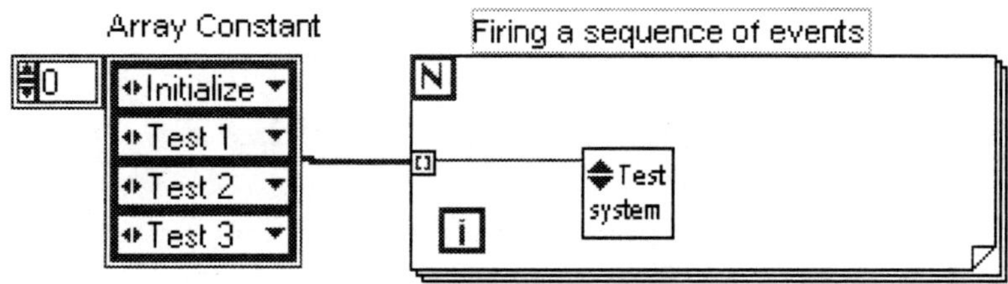

Figure 5.6
Sequence of events.

State Machine

Wiring an enumerated type into a While Loop shift register makes an extremely important structure called a state machine. The example shown in Figure 5.7 emulates a Sequence structure, but with added flexibility.

Consider the following. You've designed your component and created an enumerated type as a control and wired it to the connector. It's now deployed throughout your software. You now need to add some functionality to this component. You pop up and add a command to the enumerated type. If you put this new command anywhere but at the end of the list you will find all your popped up constants broken. If you put the new command at the end you will find ugly coercion dots everywhere. This is aggravating to say the least. The kindly people at National Instruments have given us a solution—Strict Type Definitions.

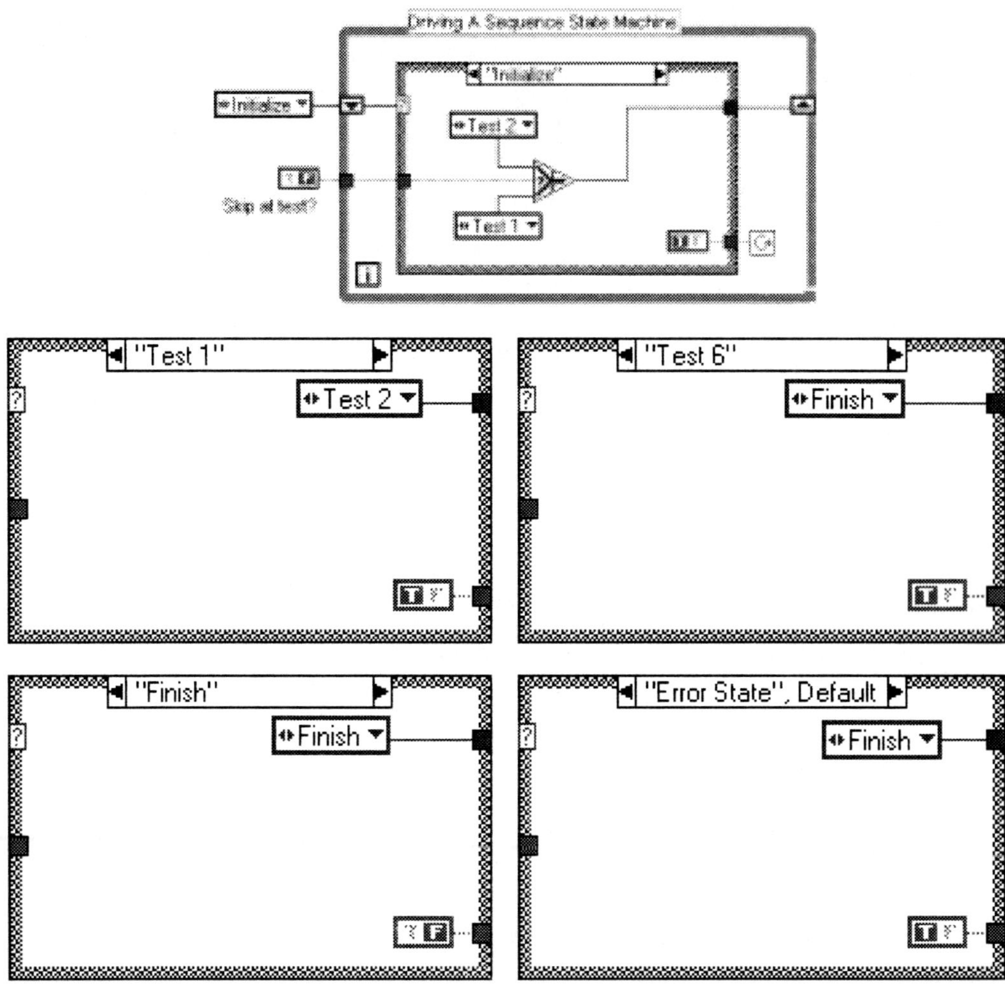

Figure 5.7
State machine structure.

5.2.3 Strict Type Definitions

You can edit your enumerated type on the Front Panel (highlight control and select Edit>>Customize Control . . ., or set your preferences to "edit a control on a double-click").

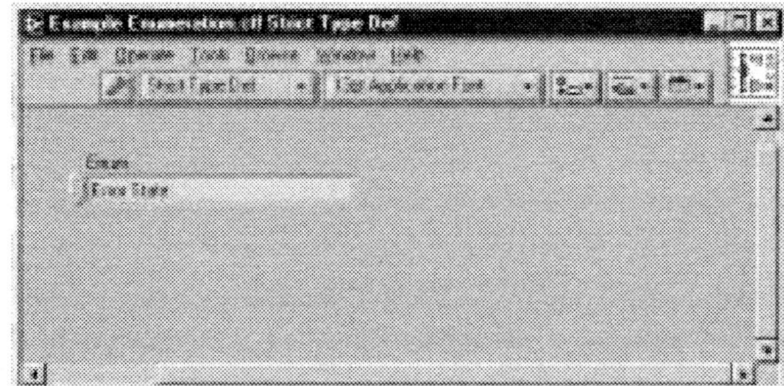

Figure 5.8
Strict Type Definition.

When the Panel shown in Figure 5.8 is displayed, select Strict Type Def. from the drop-down menu and then save in a Controls directory as *controlName*.ctl.

Now whenever you create a constant by popping up, it will be an instance of the Strict Type Definition. If you change one you change them all. Here are a few things to be aware of when dealing with Strict Type Definitions.

Bad Things

- If you create a subVI by selecting it and using the Create subVI, the new Front Panel objects are not linked to the Strict Type Definition.

- Similarly, popping up on a tunnel in a Case Structure does not create a tunnel.

- And popping up from any of the built-in functions (Scan From String, Unflatten From String, etc.) doesn't create them either.

Good Things

- Duplicating Cases and Sequences does preserve the link to the Strict Type Definitions.

- Cutting and Pasting also preserves the link to the Strict Type Definitions.

Prior to LabVIEW 6 you didn't get a Strings[] attribute for a Strict Type Defined Enumerated Type, but you did for a normal enumerated type. This was overcome by looping around each element in the enumeration and creating the element content array from that loop.

We were a bit worried about the robustness of using Strict Type Definitions, for example, if we changed one how would this ripple through our software. In practice they have proven to be remarkably immune to change, invariably coping with anything that we could throw at them. Everything discussed here about Strict Type Definitions is also applicable to normal Type Definitions.

LabVIEW has the tools to define a flexible command and attribute message sending mechanism. Oh so powerful! Oh so underutilized!

5.3   Persistent Local Storage

We want our component to keep ahold of its own state after it has been active. This data should be considered private, hidden from outside of the component and only accessible by the commands sent to the component.

There is only one efficient private method of storing data within a VI in LabVIEW and this is by using shift registers on a For Loop or While Loop structure. Figure 5.9 illustrates their use.

Shift registers are local variables that are primarily used for transferring values from one iteration of the loop to another. They have an important characteristic that we use, and that is persistence. By persistence we mean that the data held within the shift register is kept even after a VI has been run. This

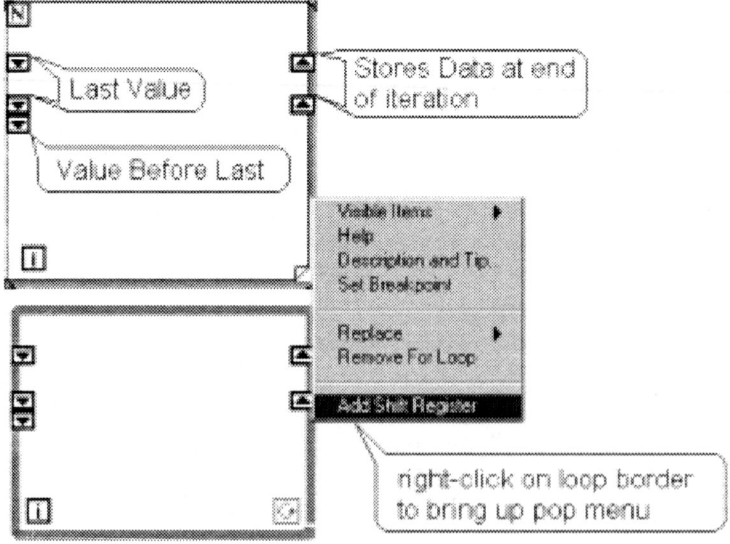

Figure 5.9
Shift registers.

gives our components independence, which is essential for loose coupling, and data privacy, which is essential for information hiding.

Let's put it all together and make a dummy component structure.

5.4   The Basic Structure of a Component

As you progress through the book you'll notice a common look and feel to all the components. We'll now build a template from scratch for a simple component structure and explain the parts.

One of the simplest structures would be an enumerated type wired to Case Structure. We'll wire the enumerated type through the shift register of a While Loop. This will allow us to write a bit of self-initializing code. Figure 5.10 illustrates this.

• The component should initialize itself.

The structure of the component allows it to auto-initialize using the very handy "First Call?" function found in the Advanced>>Synchronization Palette. The "First Call?" function is the small, round effort on the left of the While Loop. The first time a subVI is called from another VI it returns a TRUE, which causes the Initialize element of the enumerated type to be passed to the Case Structure. In the Initialize case would be file loads, array size setting, instrument setup, or any other Initialization duties you can think of. From then

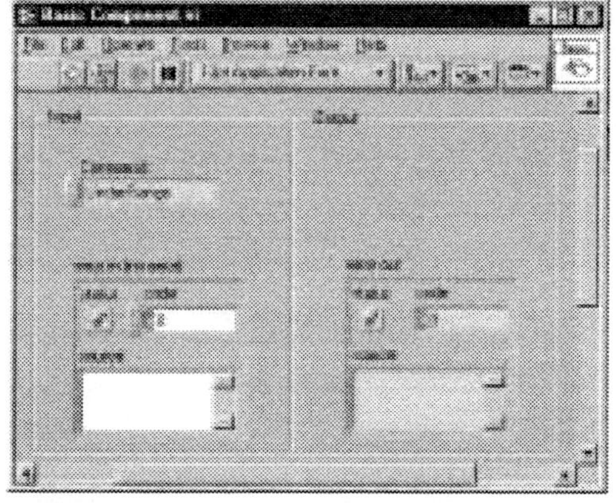

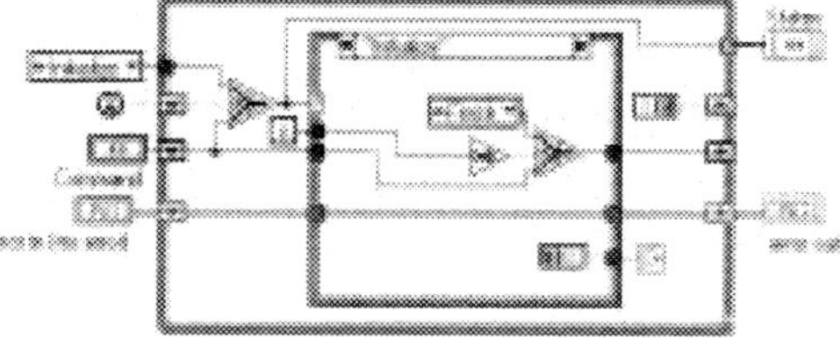

Initialize if it's the first time
through, then do the next operation.

Figure 5.10
Basic component.

on the "First Call?" function passes FALSE. Be aware that if you run the Basic Component VI from its Front Panel it will always return a "First Call?" of TRUE. The Initialize case then passes the Command enumerated type to the shift register and iterates another time.

If your component doesn't need initializing, just leave that particular command out.

- Errors are handled by the component.
- Inputs and Outputs should verify themselves.

As seen in Figure 5.11, for this template all the cases either Finish and exit or just exit. The UnderRange and OverRange Cases protect the bounds of the enumerated type, and they should be used to throw an error complaining that they've been called. For this example the error-trapping consists purely of passing the errors through. There is no reason that the error conditions shouldn't be separate cases or that part of the Finish case has the capability to check for errors and act accordingly.

The Finish case could be used for postcondition checking, and you could even employ a Start case to check for preconditions. Pre- and postconditions are described in detail in Chapter 6 on complementary techniques.

- We should be able to add/delete/modify our component's actions simply.
- Any modifications should have little effect on the overall software design.

To add new functionality to the component simply add the function name to the enumerated type and duplicate or add a new case to the Case Structure.

Figure 5.11
Component states.

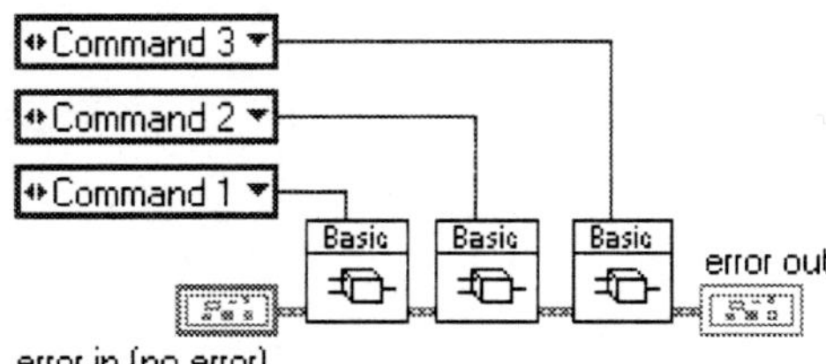

Figure 5.12
Basic component public interface.

All you need to do then is add the new actions. When deleting functionality it is worth removing the obsolete cases before removing the elements from the enumerated type.

- All of the components, public functions, and data should be accessible via a simple interface.

The public interface to the component is shown in Figure 5.12, and you must agree that it is very simple. This component is obviously very simple because it has no real inputs and outputs. That aside, if you want to access any of the functions of the component all you need to do is right-click on the Command input, and a Strictly Typed enumerated constant will be created, listing all of the functionality available for the component.

- The component stores its own state locally and persistently.

Our example only uses its shift registers to contain data when it is running; it has no need to store any local data between calls. If it did, as does the stackqueue component described later in the book, then you would see one or more shift registers initialized by the Initialize case.

LCOD Complementary Techniques

6

6.1   State Machines

Alan Turing first proposed the state machine in 1936. He envisaged a box with a paper tape input (input alphabet), a set of states (and start states), and a mechanism for mapping the input to the next state.

> **Definition**—A state machine is a conceptual machine with a given number of states; it will only be in one of the states at any given time. State transitions are changes in state caused by input events. In response to an input event the system may transition to the same or a different state, while an output event may be optionally generated.

For our purposes Enumerated Types and other assorted inputs replace the paper tape input. The mechanism for setting the start states is an initialization function, and the mechanism for mapping an input to the next state is the case structure. If the state machine needs to manage its own data you will need to include a While Loop structure and a shift register.

If you are beginning to think this is a little archaic, you could always call your state machine an object, package, module, component, or whatever.

So let's try out a simple and reasonably useless example before diving into the good stuff.

6.1.1 State Machine Example—Washing Machine 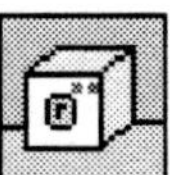

The standard example given to describe a state machine is the washing machine, so who are we to buck a trend. Figure 6.1 shows the states available to a washing machine.

Select your start state and run. It's best to view the diagram with Highlight Execution selected. If the subVIs are set up to run properly, you will find that the display light emitting diodes (LEDs) will run in order, displaying the current state of the washing machine.

Figure 6.2 is the hierarchy of subVIs, each one can be selected and its output condition chosen.

So looking at the block diagram in Figure 6.3 the start condition "Door Closed" is input into the shift register and then is passed to the case statement. The door state is returned (this could be from a switch reading) and then if the door is closed, the next state "Fill with Water" is called and so on.

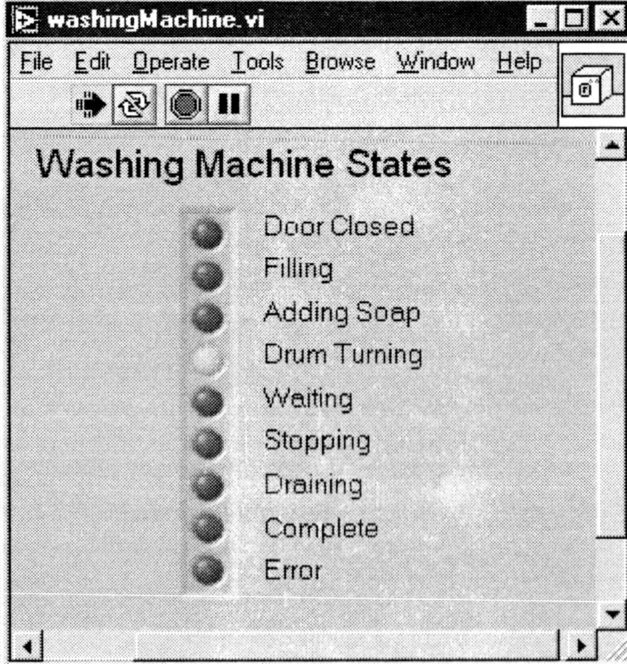

Figure 6.1
Washing machine state machine.

The other states are modeled in Figure 6.4.

If you want to experiment with the actual state responses, pop up on the relevant subVI and change the enumerated settings control.

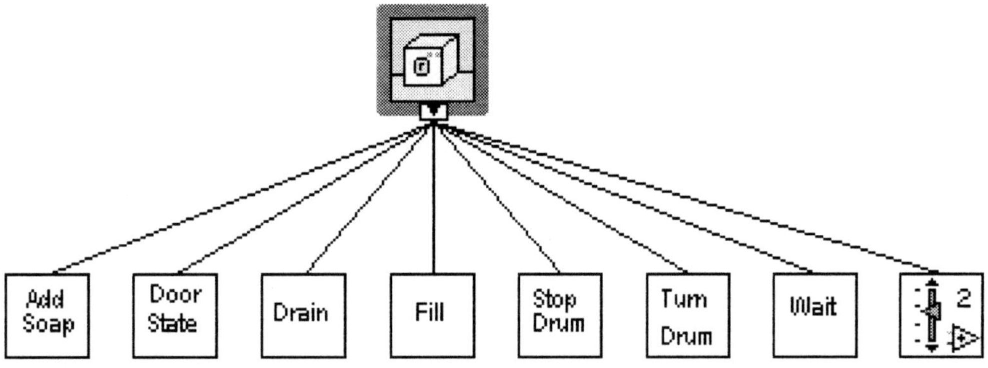

Figure 6.2
Washing machine state subVIs.

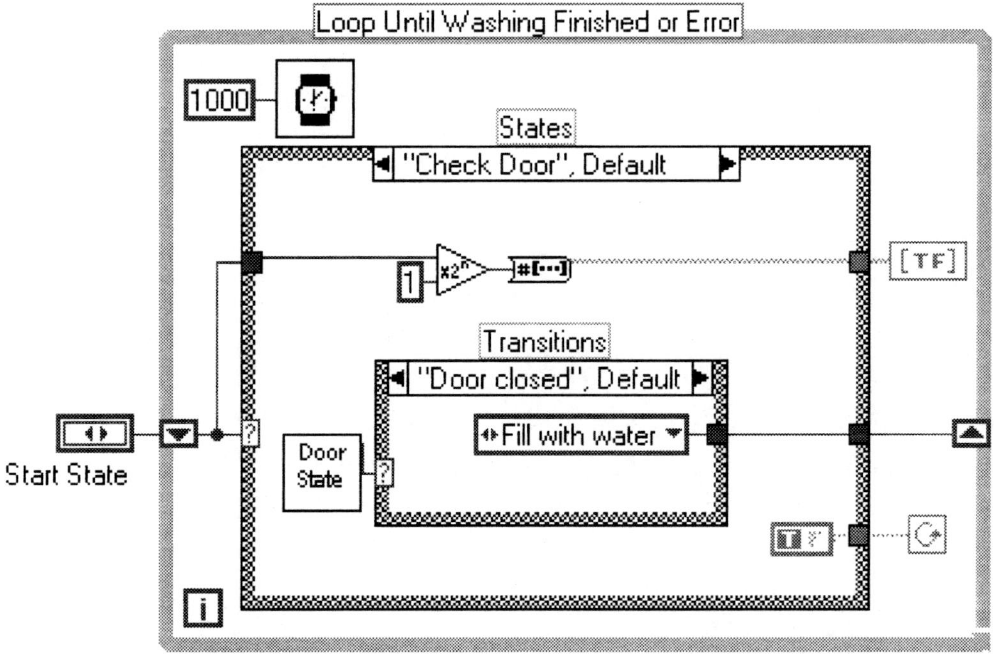

Figure 6.3
Washing machine block diagram.

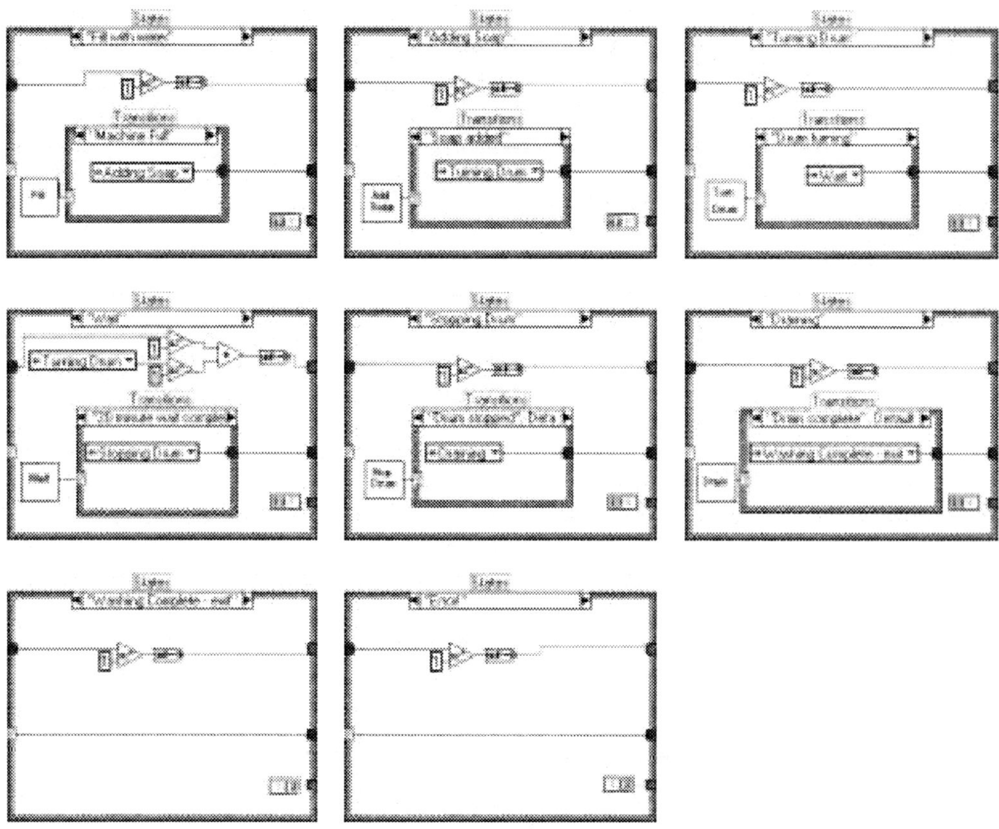

Figure 6.4
Washing machine states.

The corresponding state transition diagram is shown in Figure 6.5. The states are the boxes and the transitions are the arrows.

So why are state machines better than flowing it in the normal LabVIEW way?

The strictly sequential way in Figure 6.6 looks okay from a diagrammatic point of view, nice and simple, but what if we want to add a new state? Or change the order of states? Both of these would involve deleting the source code and moving VIs around. You don't have the options and control of flow that you get with the state machine structure.

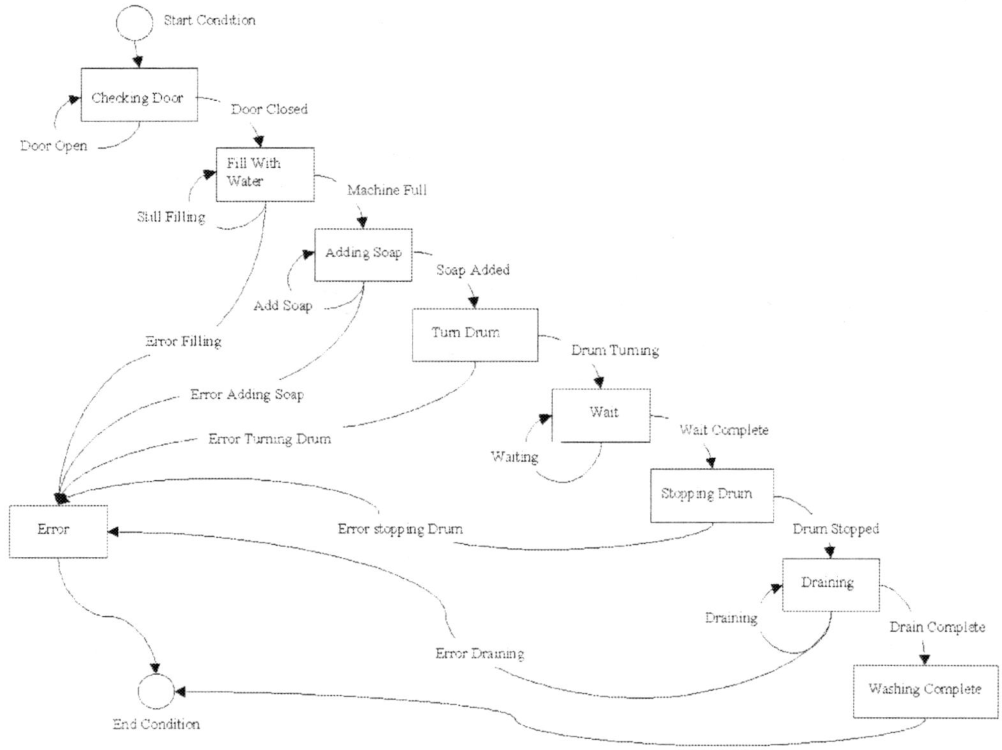

Figure 6.5
Washing machine state transition diagram.

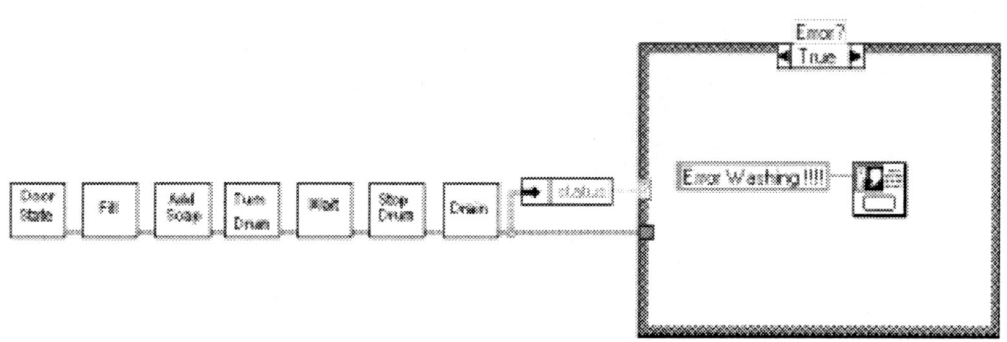

Figure 6.6
Sequential washing machine block diagram.

The error handling is also more vague. If you model the error conditions as valid states of the system you will know what error was called, and you would also know that only the actions you control would be called on an error.

State machines are an excellent method of reducing complexity in your software. They allow your code to be flexible, readable, and maintainable.

6.2 Graphical User Interface (GUI) Design and Prototyping (UI Controller>>Message Queue Pattern)

Most of us initially have problems with UI implementation and controls in LabVIEW. They can end up having large, flat block diagrams or the use of globals (elegant, but a coupling and information hiding no-no). Some LabVIEW gurus suggest using globals as the only way to program complex LabVIEW UIs. They are WRONG! This section will demonstrate how you can view your Front Panel as a container of Controls that have Attributes and Actions associated with them. It would be nice if these Attributes and Actions could be triggered from anywhere in your applications hierarchy. Well, why not pass them as messages in a queue. Then, on your Front Panel you could read these items off the queue and respond to them. This is classic message sending. In effect, you are changing your UI into a state machine and queuing its new states. Figure 6.7 demonstrates this queuing mechanism.

Let's run through an example.

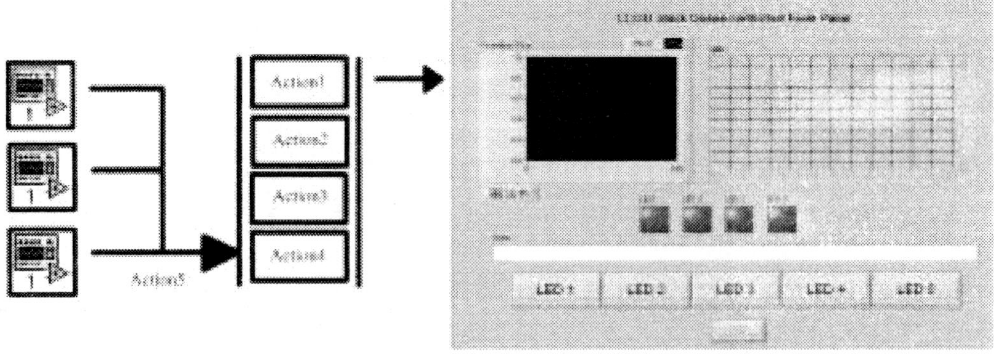

Figure 6.7
UI control example.

The Front Panel in Figure 6.8 is a simple demonstration of the principles discussed above. We will populate it with a few Indicators and Controls and then control them by queuing messages.

Either the graph or the table will be visible depending on the graph/table button condition as shown in Figure 6.8. Pressing the LED buttons will toggle the corresponding LED. In the background, the status string updates. This User Interface has minimal complexity, but you could add 10 times the complexity without changing the program structure markedly.

A list of what you want to do to it is shown in Figure 6.9.

As you can see, the UI states have already been typed into an enumerated type, allowing us to edit and remove items at will. Just pop up on the enumerated type and add the new function, its attributes, and update the UI control VI.

This is something you could sit down with your customer and define.

This enumerated type becomes the command for the wrapper VI that passes all the messages into the queue (UI Control.vi).

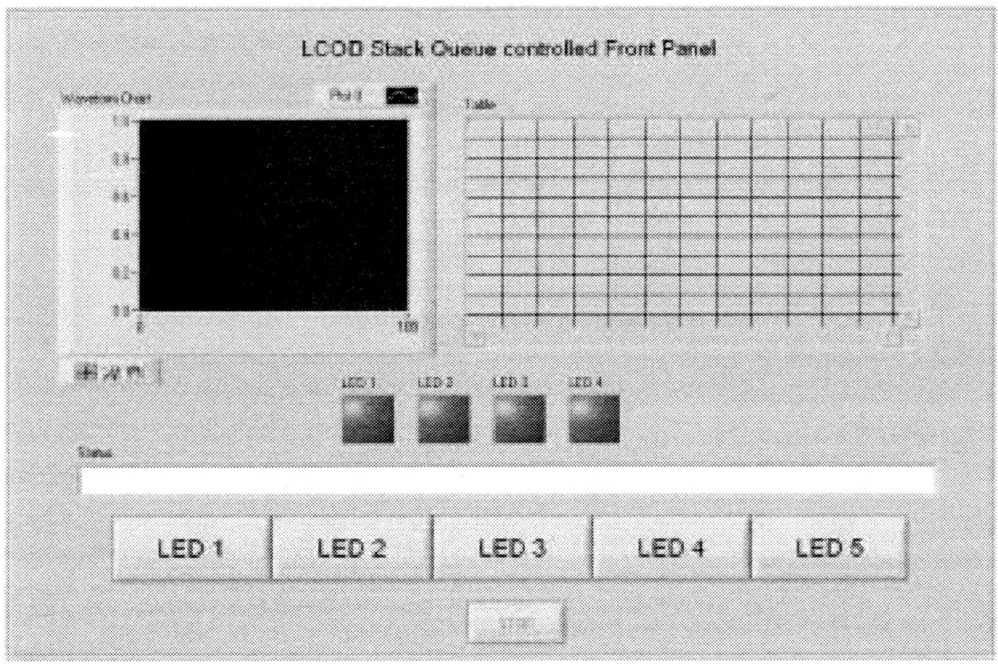

Figure 6.8
LCOD User Interface example.

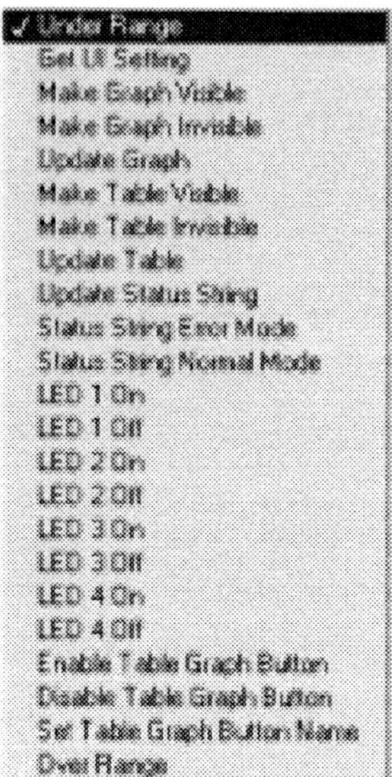

Figure 6.9
User Interface actions.

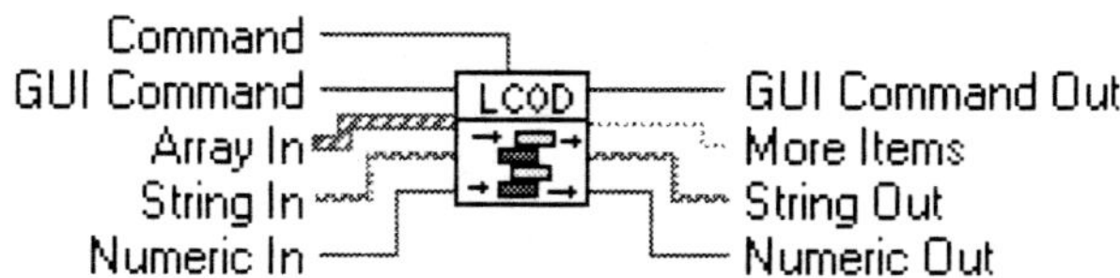

Figure 6.10
Stack queue component
connector.

6.2.1 Stack Queue Component

This little component, shown in Figure 6.10, is useful for many purposes. We tend to adapt it the most for Front Panel control, background filing, and database work. Figures 6.11, 6.12, 6.13 and 6.14 demonstrate how it functions.

There are also several useful utilities for reading the top and bottom and checking if the component is empty. Here is the implementation in LabVIEW 6.

Selecting "Add to Top" or "Add to Bottom" on the enumerated type "GUI Command" shown in Figure 6.15, will pass the data into Array Local, either at the beginning or end of the array. You then have the option of taking this data from the beginning or end of the array with the commands "Pop from Top" or "Pop from Bottom". Figures 6.16 and 6.17 show its implementation.

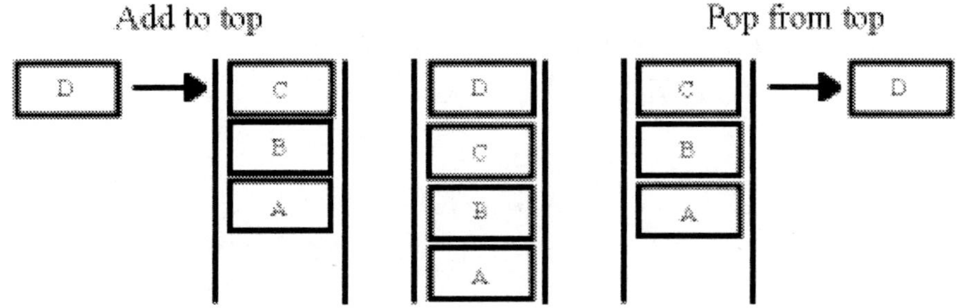

Figure 6.11
Stack.

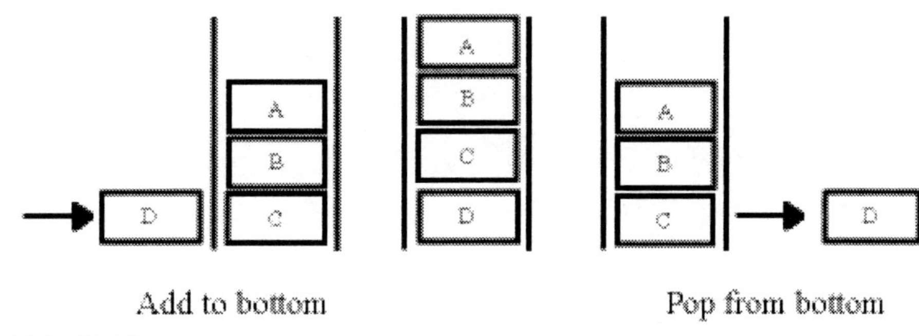

Figure 6.12
Upside-down stack.

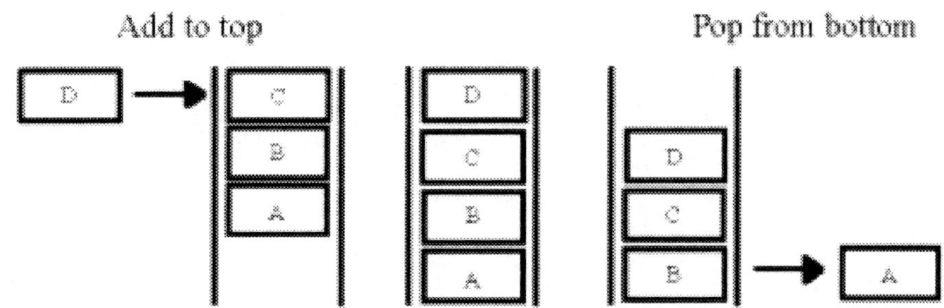

Figure 6.13
Queue.

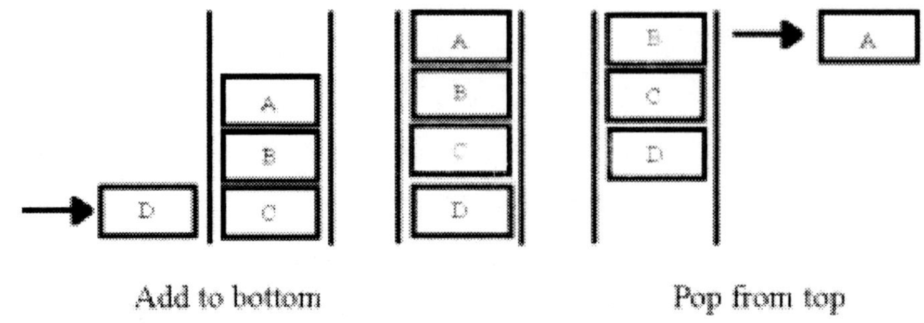

Figure 6.14
Upside-down queue.

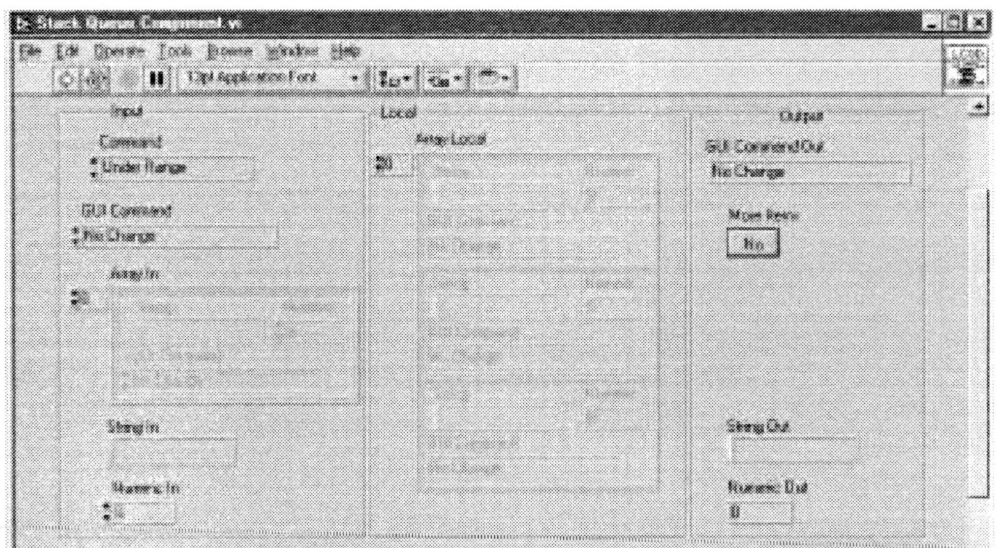

Figure 6.15
Stack queue Front Panel.

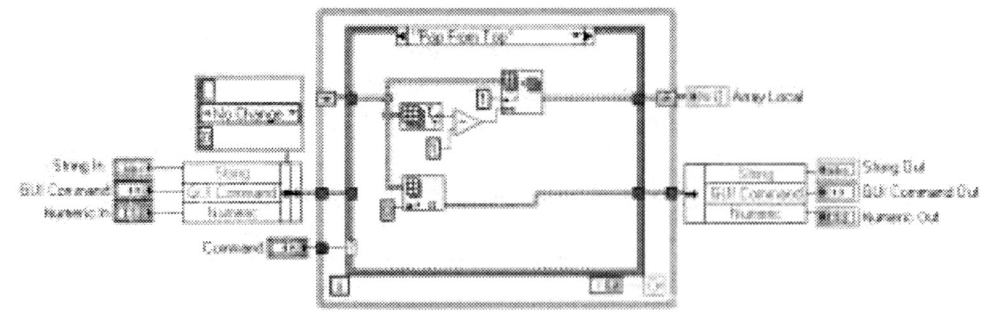

Figure 6.16
Stack queue diagram.

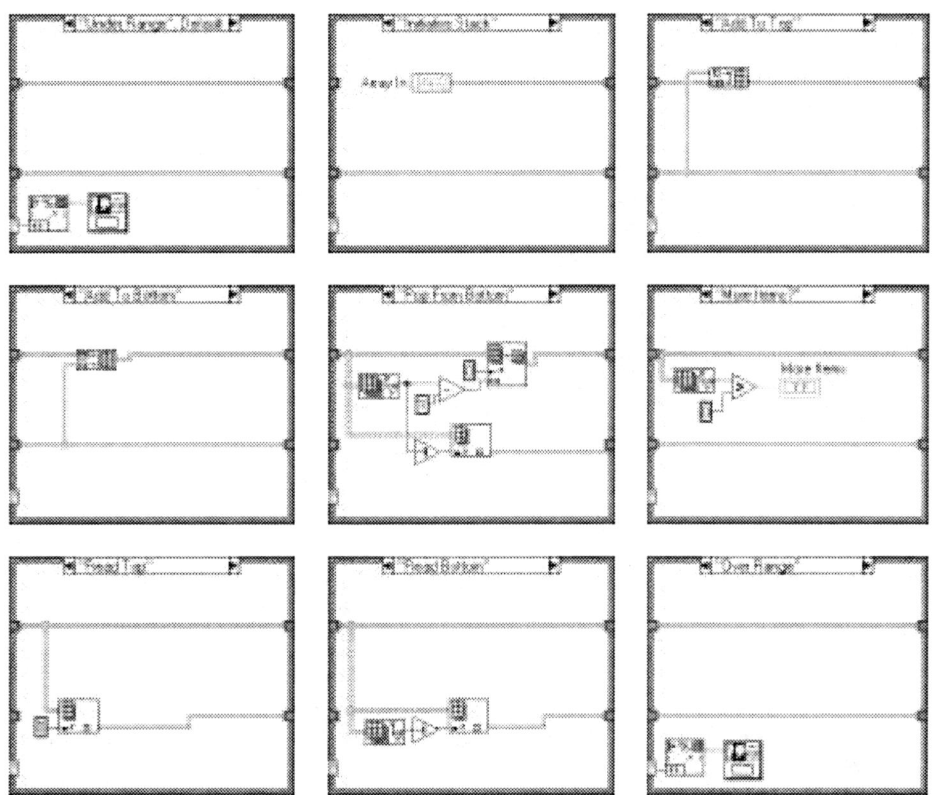

Figure 6.17
Stack queue cases.

It's relatively easy to modify the component to your own data requirements by changing the inputs, clusters, and array contents to suit. Make sure that the source code is left the same though.

We'll use this component to store and transmit the states of the User Interface.

The stack queue component works fine for small queues, but if you want to beef up the performance you could try using the queue VIs that come with LabVIEW. In LabVIEW 6.1, NI very kindly made them polymorphic and they work nicely, giving a significant performance improvement. We rewrote the component to include these VIs and they were about eight times as fast for large queues (2,000+ elements).

How to get to the queue functions can be found in Figure 6.18.

We've kept the functionality and the interface similar, although there are now error clusters in and out. You can only take items from the end of the

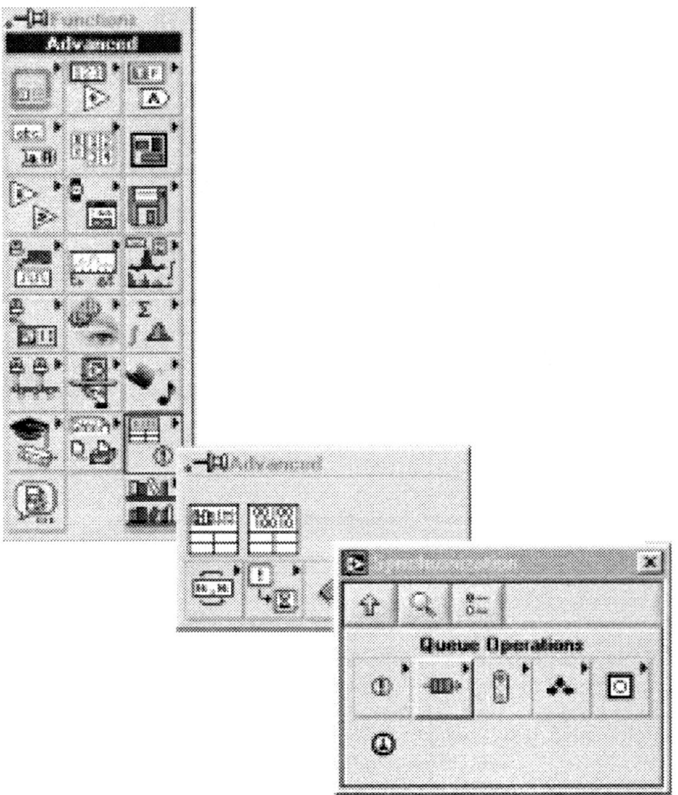

Figure 6.18
Queue operations.

queue—a small sacrifice for the speed advantage. The element cluster is a strict type def, therefore, to reuse the queue all you have to do is customize the element cluster. This can be done using the queue VIs in earlier versions of LabVIEW, but you have to squish the data to a string and unsquish it afterwards. Also you lose the ability to use the component like a stack, the queue functions constrain you to queuing in one end and out the other.

The source code for this implementation is in Figure 6.19.

6.2.2 User Interface Control Wrapper VI

Next you have to create the wrapper component that conducts all the actions on the stack. For improved maintenance you could map the input command

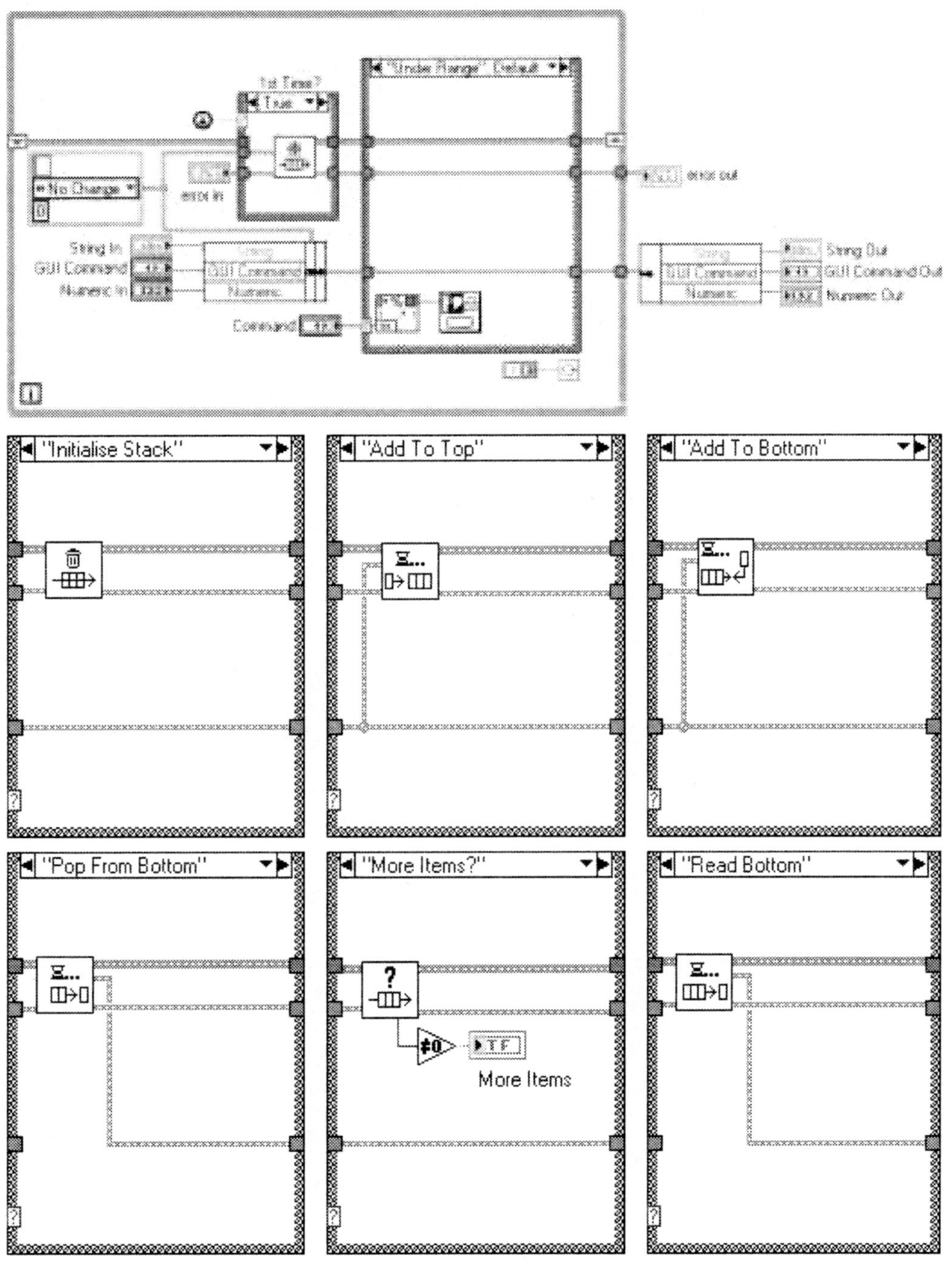

Figure 6.19
LV6.1 stack queue component diagram.

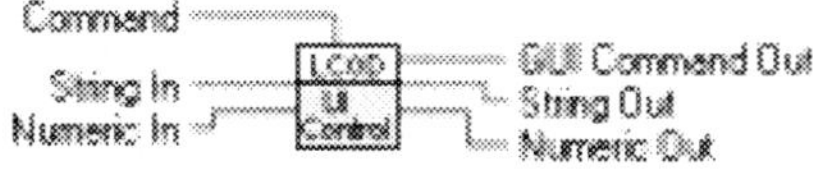
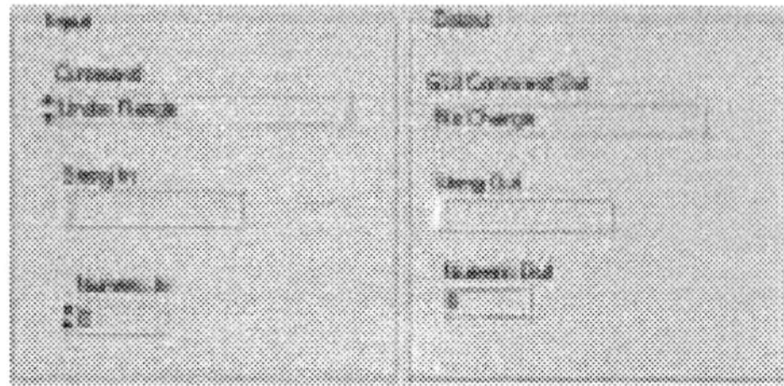

Figure 6.20
UI Control.vi.

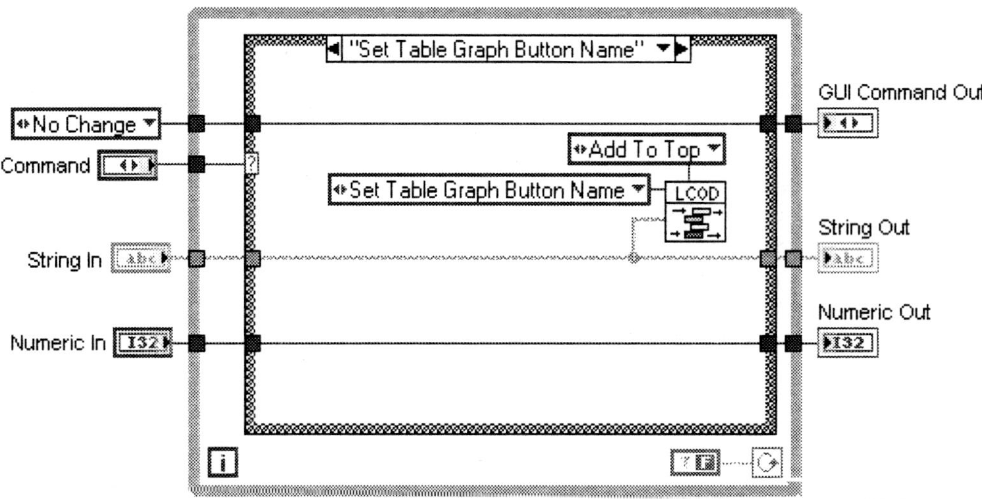

Figure 6.21
UI control diagram.

directly to the "GUI Command Out", but this has a negative effect on the flexibility of the design. Figures 6.20, 6.21, and 6.22 describe the implementation for this VI.

Looking at Figure 6.21 you can see the case for the display state, "Set Table Graph Button Name". When this command is selected and the VI executed the corresponding command and string are placed in the queue. A loop in the UI block diagram is then responsible for pulling the commands and settings off the queue and executing them. The case, "Get UI Setting" in Figure 6.22 is responsible for this.

Adding a new User Interface attribute or action is merely a matter of updating the enumerated type, duplicating the case, and changing the enumerated type in the case to suit.

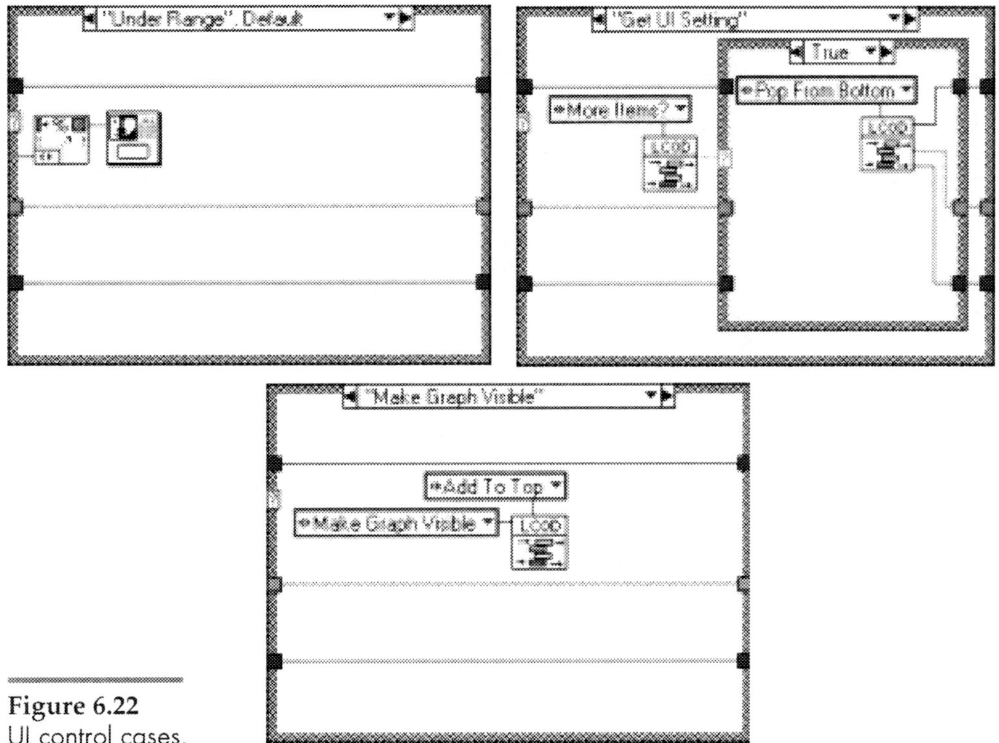

Figure 6.22
UI control cases.

6.2.3 LCOD User Interface Example Diagram

Now put in the state machines for monitoring User Interface updates and
checking for button presses. You will also notice in Figure 6.23 that everything
is enclosed in another state machine that describes the states for the User
Interface (initialize, run, finish).

You then fill out the "Monitor for UI Changes" state machine for all of the
attribute settings and actions as shown in Figure 6.24.

As you can see, all the attributes and actions for the User Interface are held in
a nicely documented case structure. Adding more functionality is just a matter
of creating another pigeonhole and placing the control or property node into it.

You can also make a small test stub VI that pushes actions onto the queue,
allowing you to test each individual display setting. Just drop the UI Control
into a new VI and set each input as a control. Run your User Interface and on
the test stub select the action you require and run the test stub. As if by magic,
the UI will do your bidding.

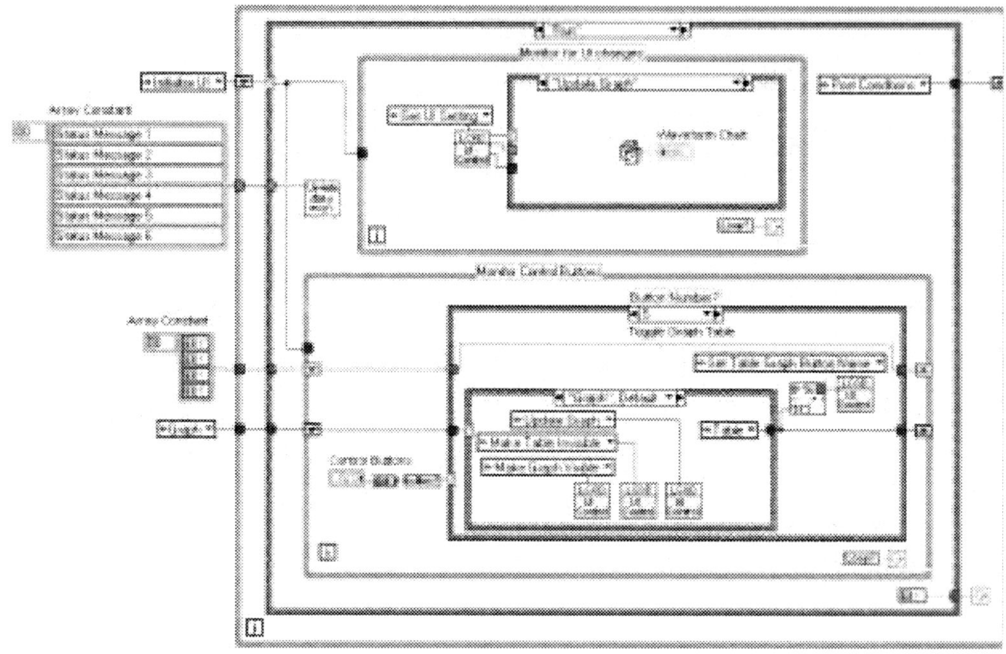

Figure 6.23
LCOD Front Panel diagram.

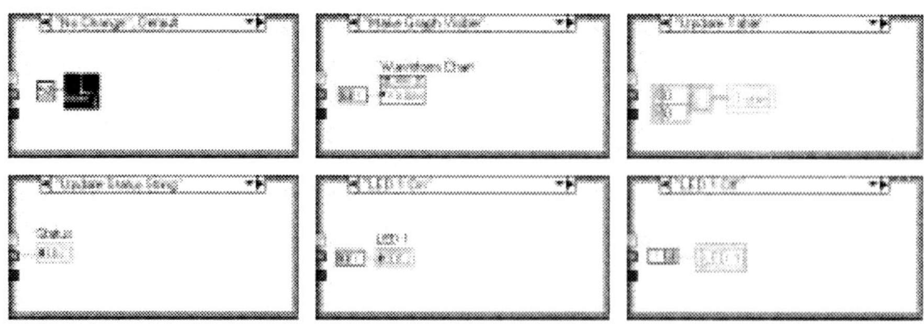

Figure 6.24
UI attributes and actions.

6.3 Abstraction in the Code, Detail Outside the Code

What does "Abstraction in the code, detail outside the code" mean? Well, essentially the software should encapsulate the services that are to be provided, but there should not be any detail in the code. What?

To make this clearer we'll define what we mean by detail. Detail would be a GPIB address, measurement parameters, switching connections, and so on. The abstraction would be the component that provides the interface to the instrument at that address, the component that commands the instrument that makes the measurement, and the component that commands the switching.

Here is an interesting fact that we have noticed (that it is something we have noticed probably means it is not a fact, but go with it). When software is beyond the initial construction stages, it's the strings and integers that are changed more than the actual code. This is a double-edged sword because if you design the software with this rule in mind, you also produce code that is better designed, more flexible, more reusable, loosely coupled, and hence, easier to maintain.

Okay, this sounds a bit soapbox, so here is the golden rule:

When you look at your VIs, every time you see a number or string it should be taken out and placed in a file or database, yes, every single one of them!

We're serious. Think about it, every string and number in your code is detail, and detail changes. The GPIB address changes, the frequency span of your measurement changes, and the switching changes with a design change or addition of new hardware—these are all detail. But, the abstraction doesn't change. The instrument can be at any address, the address does not change how the component operates, a frequency measurement is the same regardless of the span, and switching is the same operation regardless of the location. As a simple illustration, your program is compiled to an exe and the hardware GPIB address has to change because of a clash somewhere, all you have to do is change the detail in a file (maybe a configuration file), no code changes, no recompiling. At the other extreme, the GPIB addresses are spread through the four corners of your code. If you change the address, then you have to hunt down all the occurrences—miss one and your application will fall over!

So, here's an example. Your application uses an analogue I/O board, and you want to measure data coming in on the analogue input (AI) channel. Now we have a nicely designed component that measures what we want, as shown in Figure 6.25.

But, we can see that the VI has DETAIL in it! The detail is the channel limits (array of clusters 5/−5), device (1), and channels. If for any reason this changes we have to change the code, which is exactly what we want to avoid.

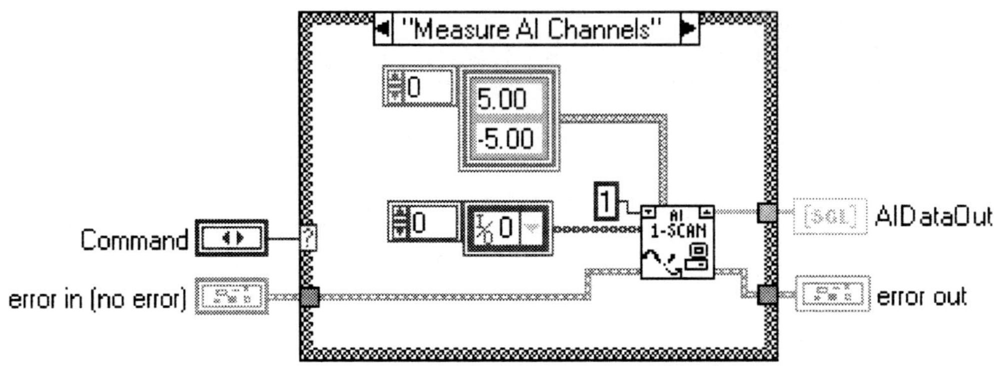

Figure 6.25
Detail in the code—Bad!

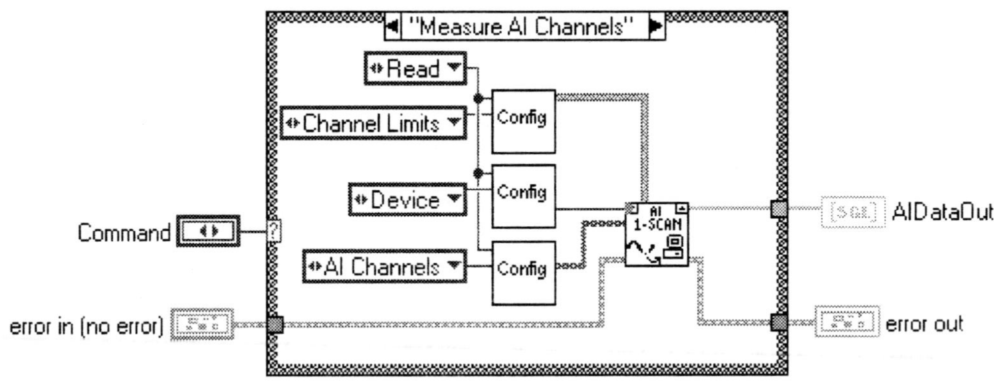

Figure 6.26
Detail outside the code—Good!

So, an updated version is shown in Figure 6.26.

We can see that there are no longer any constants in our code, they have been replaced by three instances of the VI called Config. Simply put, the Config VI has the detail stored in a file outside the code.

As with most things in this book this seems simple. So, all detail in files or databases, okay that's easy. But bear in mind there will be huge dividends gained by adhering to this general rule. However, the simple thing to do is put the detail in the code with the mental note that you will take it out later (we know this because we're tempted every time). Multiply this by 500 VIs and you've missed the boat. Therefore, when you recognize detail, stop and put it outside the code. In the short term this becomes a little monotonous, but soon most of the data will be extracted and you will just reference it in later VIs.

6.3.1 Section Key Files

You will need a place to store all of the detail data previously discussed, and we've found an ideal solution in section keyed files. There is a set of useful functions available on the function palette for implementing these files, as shown in Figure 6.27.

What is a section key file? It's the old .ini file as used by Windows 3.1, and it is a Text file with the following format:

```
[section name]
key1=Data1_1|Data1_2|Data1_3
key2=Data2_1
key3=Data3_1
```

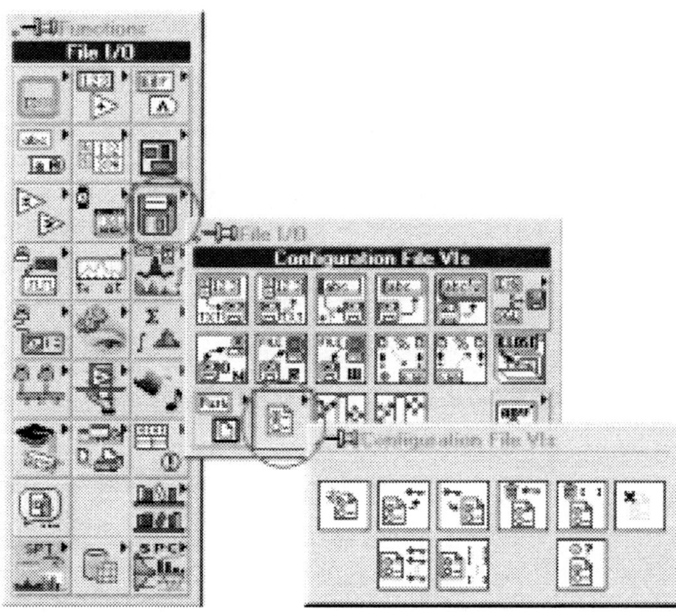

Figure 6.27
Configuration file functions.

The data held in these files is essential to the running of your application, therefore, there is an issue of security to consider, we don't want anyone coming and idly fiddling with the data held in these files. The configuration file functions as they stand do not give any encryption facilities, so as a first step let's add some.

All we need to do is replace "Config Data Read From File.vi" and "Config Data Write To File.vi" in "Close Config Data.vi" and "Open Config Data.vi" with encrypted versions and save them under a different name.

Let's do it step by step.

First open "Open Config Data.vi" from the functions palette and save it as "Encrypted Open Config Data.vi" as in Figure 6.28.

Open up the block diagram and double-click on "Config Data Read From File.vi", as shown in Figure 6.29 and Figure 6.30.

Figure 6.28
Open Config Data icon.

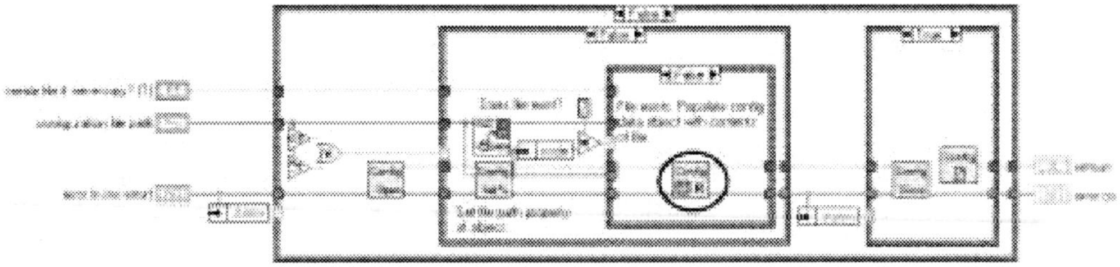

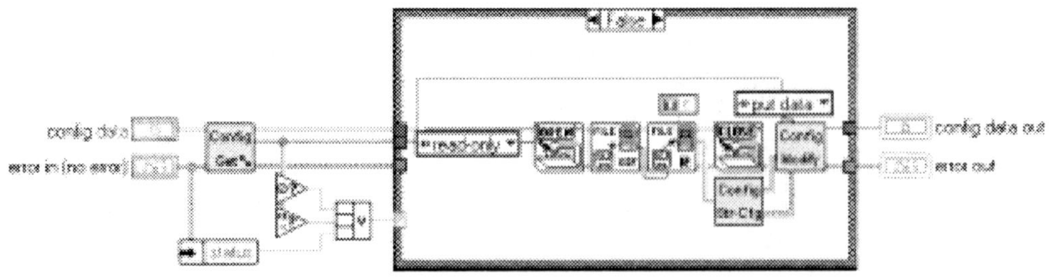

Figure 6.29
Open Config Data diagram.

Figure 6.30
Config Data Read From File before.

Rip out all of the file opening and reading stuff and replace it in Figure 6.31.

So, we read the file as a bunch of integers and push them through an encryption component that decodes them. The encryption VI does some rotating and math on the incoming integers.

That's the data decoded when we read it. Now, let's finish it off by encrypting the file. If all the data is read when the config file is opened it stands to reason that it is saved when the file is closed. In a similar fashion to what we have just done, open "Close Config Data.vi", as shown in Figure 6.32. Then, double-click on "Config Data Write To File.vi", as shown in Figure 6.33.

Once again replace all the writing to file stuff with the encryption component as in Figure 6.34.

If you don't want to go through all of this you could always swipe the code off the Web site.

It would be ideal if we could add items to these files as we go along, developing our application without incurring a large programming overhead. What we need is a component that handles all of this for us.

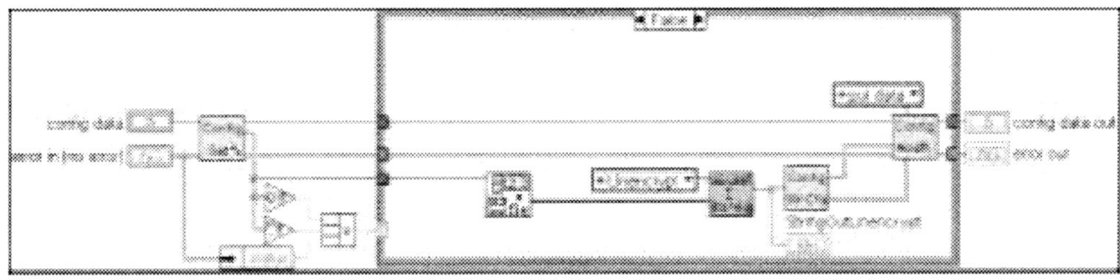

Figure 6.31
Config Data Read From File after.

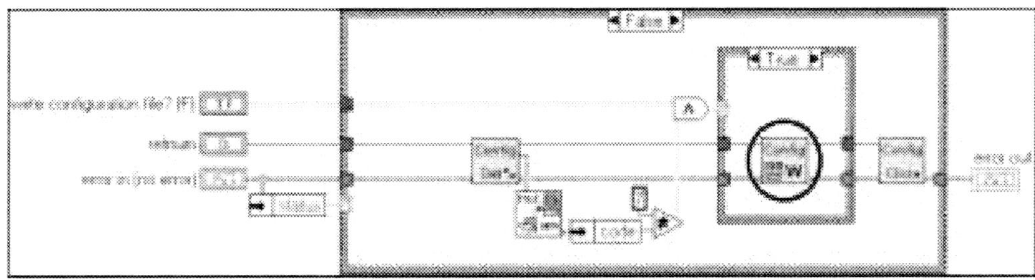

Figure 6.32
Close Config Data diagram.

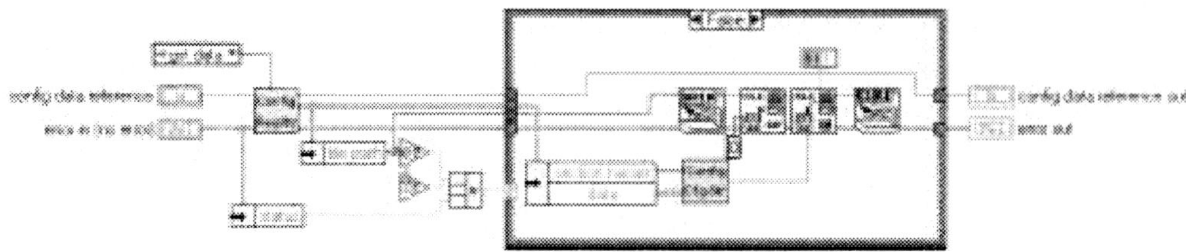

Figure 6.33
Config Data Write To File before.

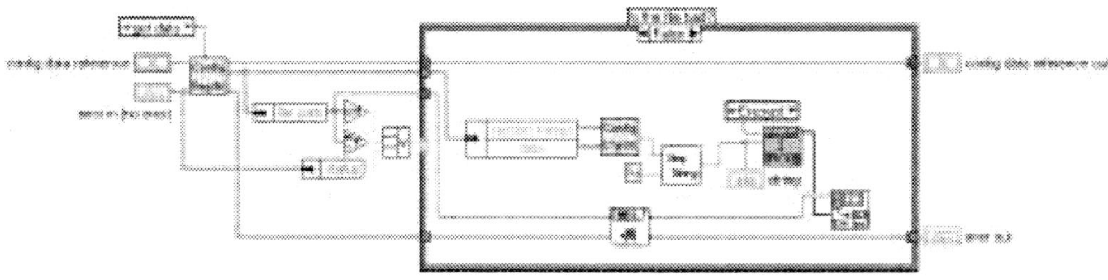

Figure 6.34
Config Data Write To File after.

By using Strict Type Def enumerations as keys and another Strict Type Def enumeration as the section, the following component allows items to be added simply by updating the relevant enumerated type. It encapsulates all of the functionality required for loading, saving, and editing data in section key files.

So, here's a list of the functionality.

Action	Description
Invalid:	This is the default condition and should throw an error
Get:	Returns the selected data item
Set:	Sets the selected data item
Save:	Saves the selected data item to file
Load:	Loads the selected data item from file
Save All:	Saves everything to file
Load All:	Loads everything from file
Reset:	Clears out local memory manually

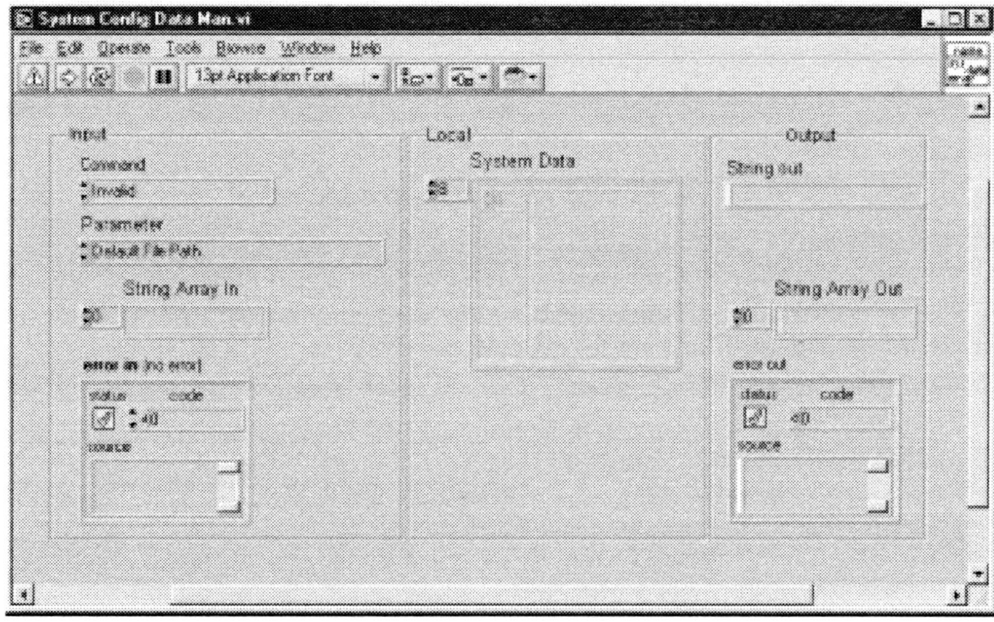

Figure 6.35
System Config Data Manager Front Panel.

Storing the data in a shift register that is an array of clusters of arrays (?!?) works okay for reasonable amounts of data, and is simple to implement. For extravagant amounts of data you could store the data in a hashed array structure, or something similar. The interface stays the same no matter what method of organizing the data is employed (demonstrating the power of information hiding).

Let's briefly discuss each of the component actions and look at the implementation.

The Front Panel is shown in Figure 6.35.

To run the component, select the action [Command] and the attribute [Parameter], fill out the data in [String Array In] if required, and run it. There is a File Handler Component that will need setting up prior to filing operations, but this will need to be done only once. It holds the references to the systems files. We'll now describe the source code in detail.

Invalid

The Invalid state, in Figure 6.36, should never be called and is only there to pick up development errors and mis-wires. If it does occur the software should throw a tantrum about it.

The code in the top left-hand part of the While Loop checks if the local array has been initialized, and if not, sets it to the number of parameters in the "Parameter" Enumeration. The two small cases open and close the files if the Command is file related. The middle case is where all the action is, and for the Invalid state it fires the Error Control into action, throwing up a dialog box moaning about being called and listing who called it.

Get

Figure 6.37 shows the "Get" action. The relevant array is selected by using the index of the enumerated type, and output as a string and an array.

Set

Figure 6.38 shows the "Set" action. Once again the new array is input at the index desired by the enumerated type "Parameter". This new data is persistently stored in the shift register.

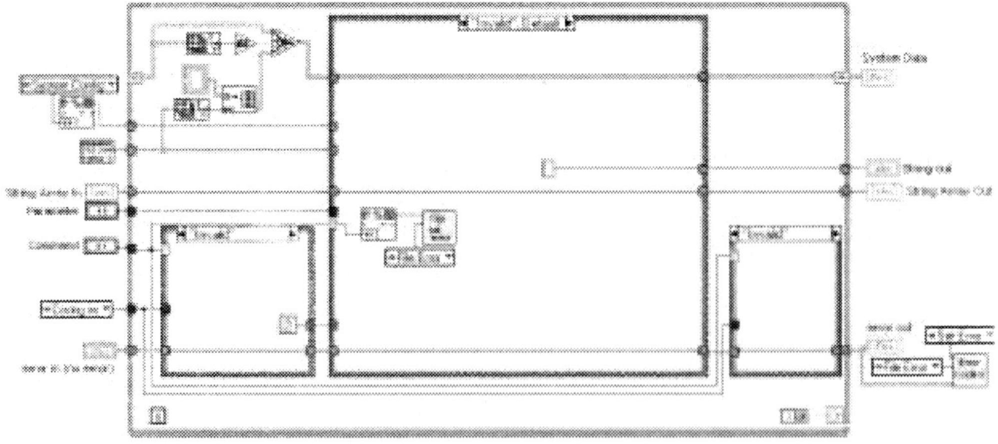

Figure 6.36
System Config Data Manager Invalid diagram.

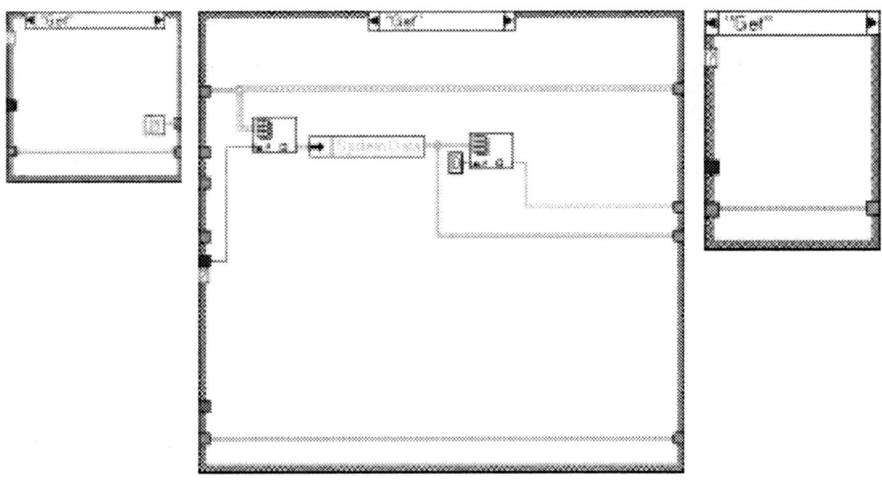

Figure 6.37
System Config Data Manager Get cases.

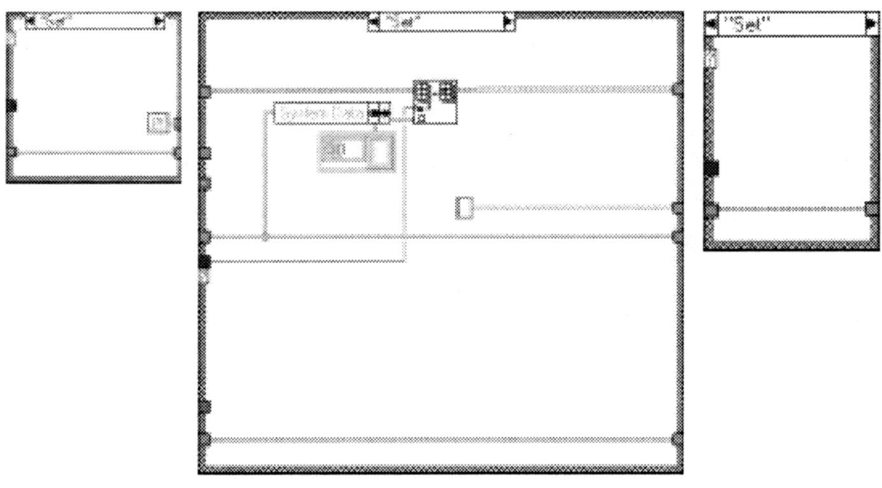

Figure 6.38
System Config Data Manager Set cases.

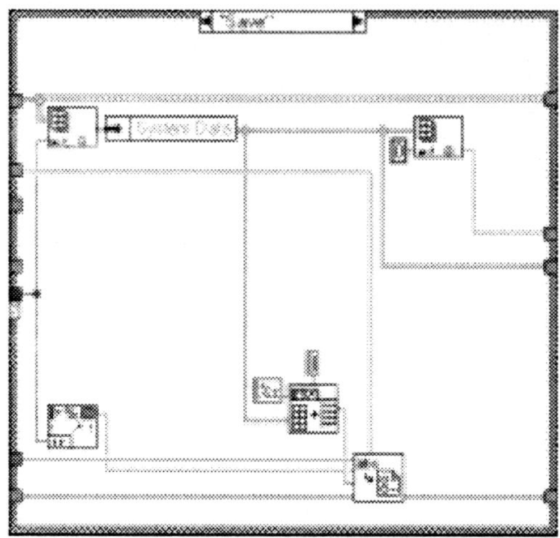

Figure 6.39
System Config Data Manager Save case.

Save

Figure 6.39 is the "Save" action. The array selected by the index of the "Parameter" enumerated type is stored in the section-keyed file at the key defined by the text of the "Parameter" enumerated type.

Load

In addition to saving the data you will want to load it from the file as well. Figure 6.40 shows the "Load" action. The text of the "Parameter" enumerated type is again used as the key for selecting the data to pull from the file. The returned string is then converted to an array and inserted into the shift register at a location defined by the index of "Parameter" enumerated type.

Save All

Figure 6.41 demonstrates the "Save All" action. The "For Loop" will auto-index for every element in the shift register array, saving to the key produced from the "Strings[]" property of the "Parameter" enumerated type.

LabVIEW 5.1 does not have the "Strings[]" property for Strict Type Definitions and this has to be done manually.

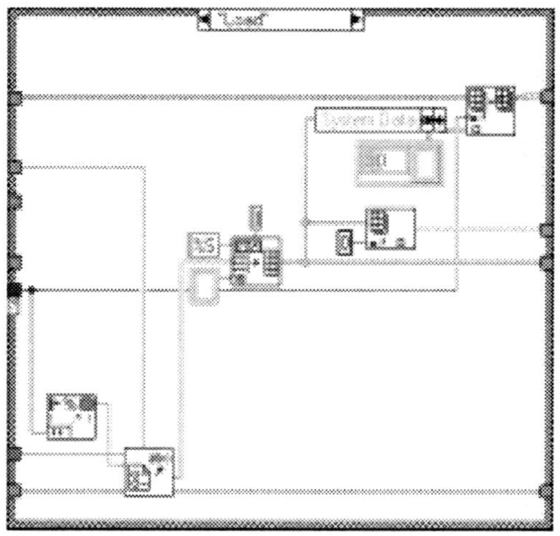

Figure 6.40
System Config Data Manager Load case.

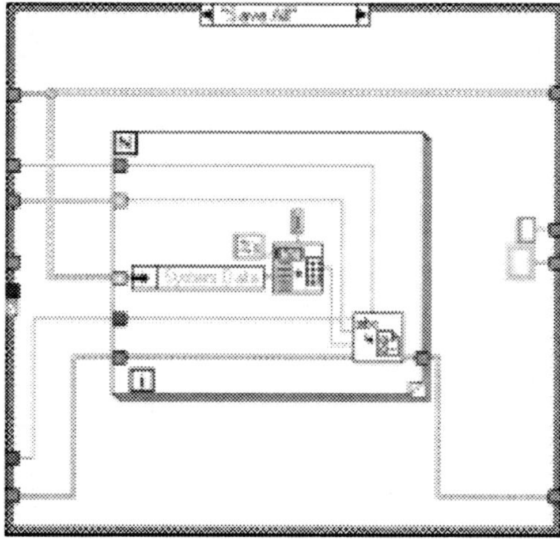

Figure 6.41
System Config Data Manager Save All case.

Load All

The "Load All" action shown in Figure 6.42 iterates through all the parameters, filling up the shift register array. The "File Handler" component holds all the file references and is also responsible for opening and closing the files.

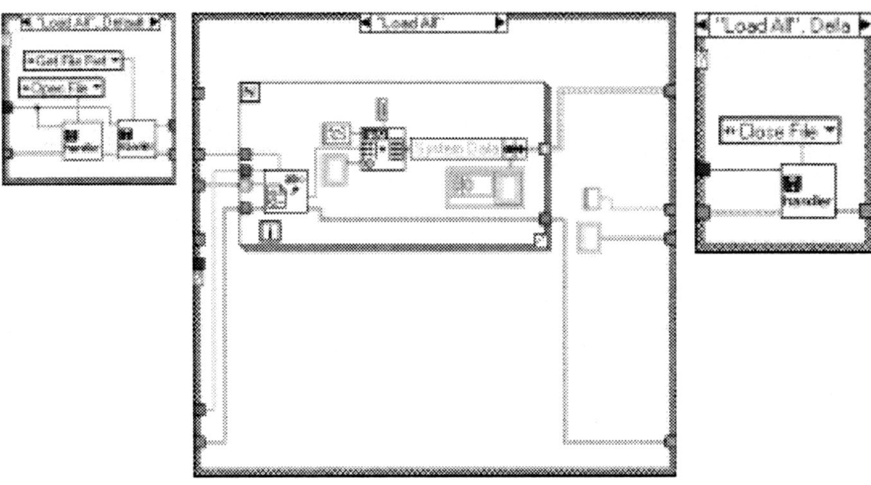

Figure 6.42
System Config Data Manager Load All cases.

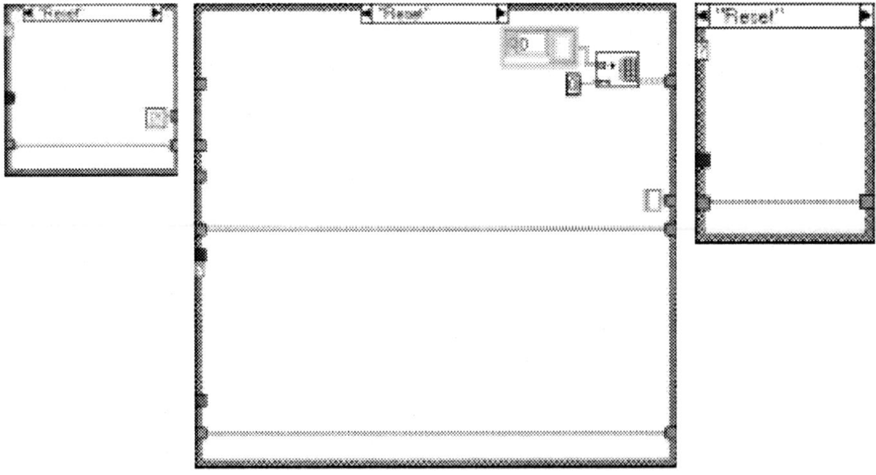

Figure 6.43
System Config Data Manager Reset cases.

Reset

The action in Figure 6.43 clears all the data held in the shift register.

If you want to have multiple data managers for storing different data all you need to do is save the data manager under a different name, replace the

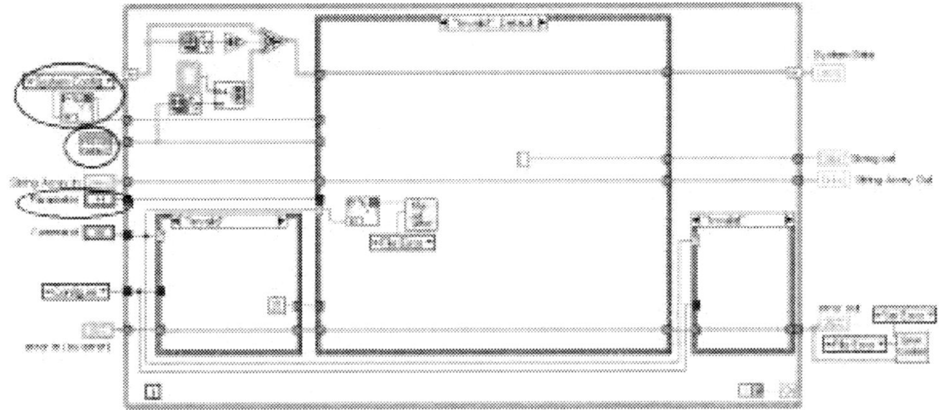

Figure 6.44
New data manager changes.

Parameter Strict Type Def, the Parameter Strings Attrib VI, and add a new element to the Section enumeration, as shown in Figure 6.44.

Finally, after doing all of this it would be useful to have a dialog that will help you manage this data. The following dialog gives full access to every piece of data held in the data managers, although this may not be appropriate, and we've seriously thought about giving some extra protection to specific data.

Once again, by using Strict Type Def enumerated types we can make the whole exercise reusable.

The enumerated type that describes the sections will be used to select the relevant data manager for display. It then will display all the keys and their data in a table. The Front Panel is shown in Figure 6.45.

Embedding this utility in your program will enable you to pull out all of your constants and manage them simply and efficiently. To update the values in the table, press [Set] to store them in local memory, press [Save All] to store everything to file, [Load All] reloads the data from file, [Print] prints the contents of the file, and [Done] closes the dialog.

Briefly the dialog initializes and then monitors the buttons. The file reference should have been set up prior to it being called, but to make a stand-alone example we've put the file reference setup as part of the initialize screen. The diagram is shown in Figure 6.46.

You can see general screen setup code (B), a table auto setup case (A), and a case that sets the file references (C).

Regarding the block diagram in Figure 6.47, we can see that the lower While Loop monitors the control button cluster, a press on any button will

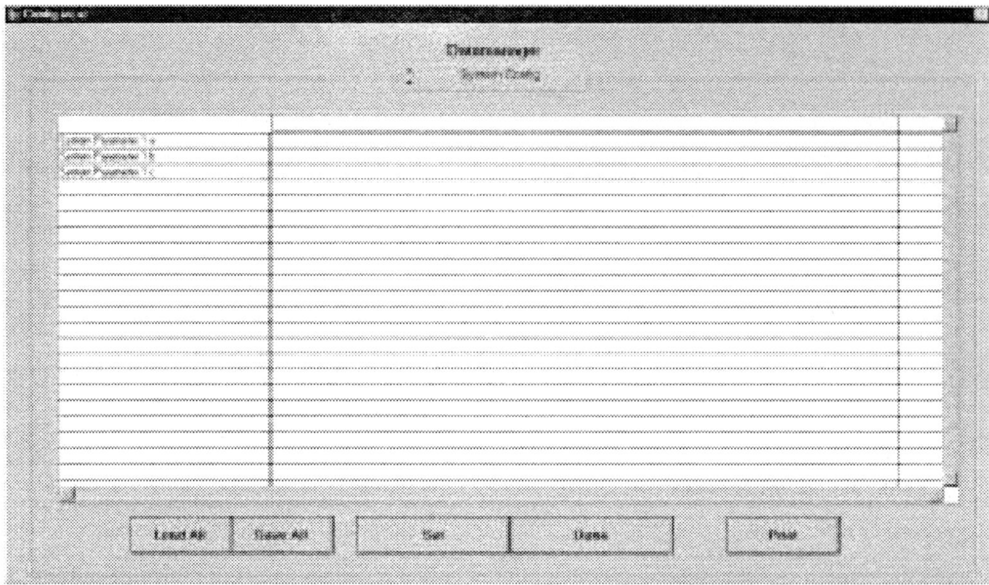

Figure 6.45
System Config Dialog Front Panel.

squirt a number into the case. The upper While Loop looks for a change in the Datamanager enumeration and loads the table on a change.

6.4 Error Handling

Decent error handling is one of the most challenging aspects of software design and, coincidentally, one of the most ignored.

When thinking about error handling there are two phases of a project that need considering: the development phase and the released phase. Each phase has separate error handling requirements. In development you want your application to bleat loudly and generally get itself noticed the moment an error occurs. You can then fault-find and fix it. When your software is released you will have to deal with errors more elegantly. For example, it will be important to have a log of the errors.

Another thing is propagation and handling of errors. LabVIEW provides the error cluster that links VIs together; any errors upstream are handled by

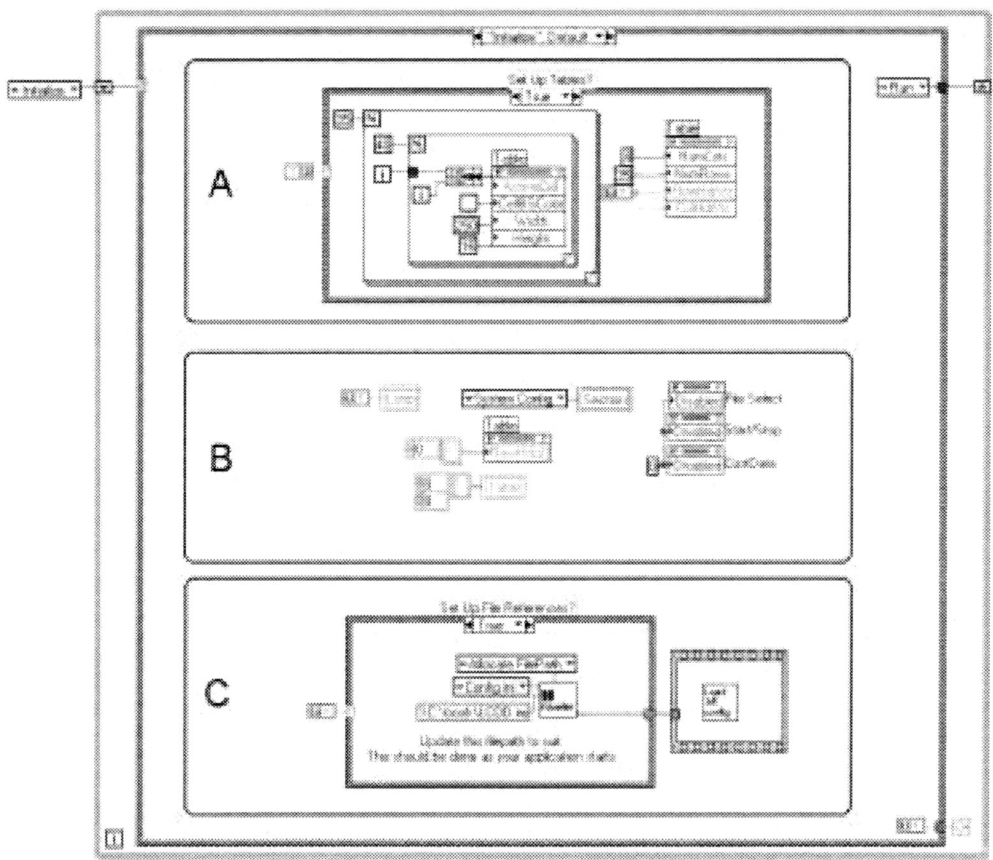

Figure 6.46
Config Ini Dialog Diagram Initialize.

the next VI in the sequence. There are questions of coupling with this approach, for example, does the next VI need or want to know about errors in its predecessor?

Consider a system designed using the standard data flow method as shown in Figure 6.48.

If an error enters VI number 2 it will pass through without completing its function—this will be repeated for VI numbers 3 to 6. LabVIEW is a parallel processing, multithreaded language. So the simple data flow model seldom holds true. So, for example, if you have multiple While Loops processing in

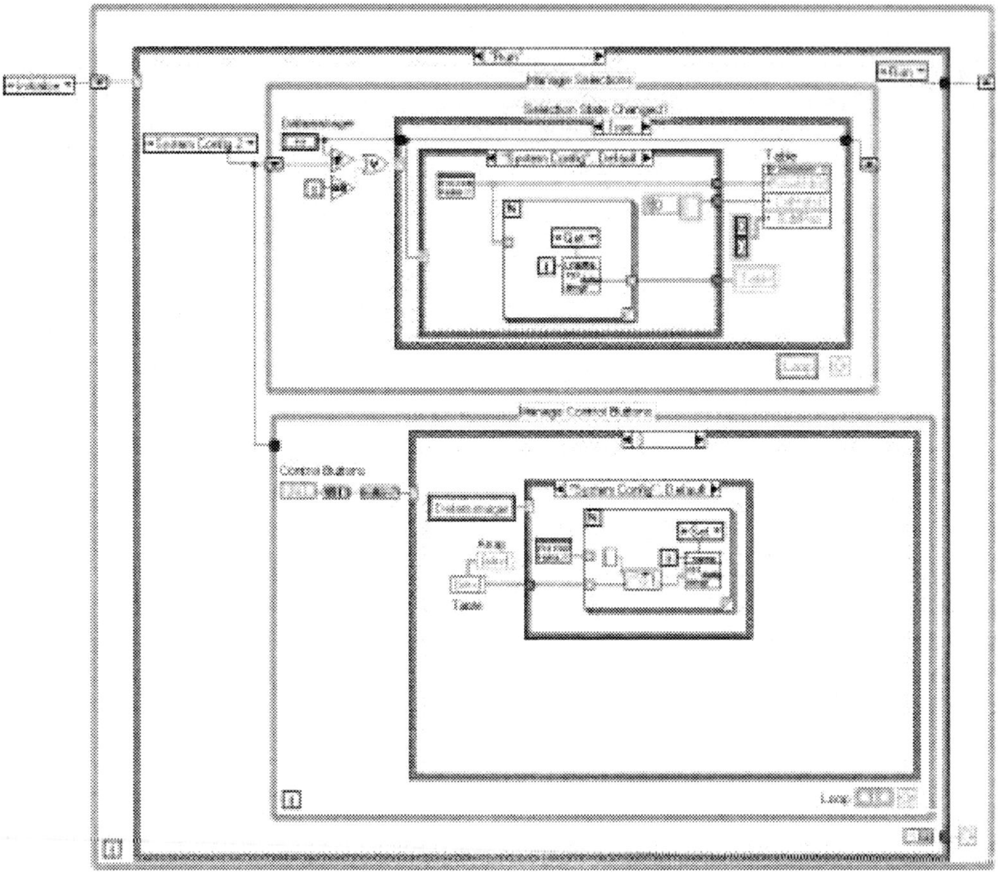

Figure 6.47
Config ini Dialog Run State.

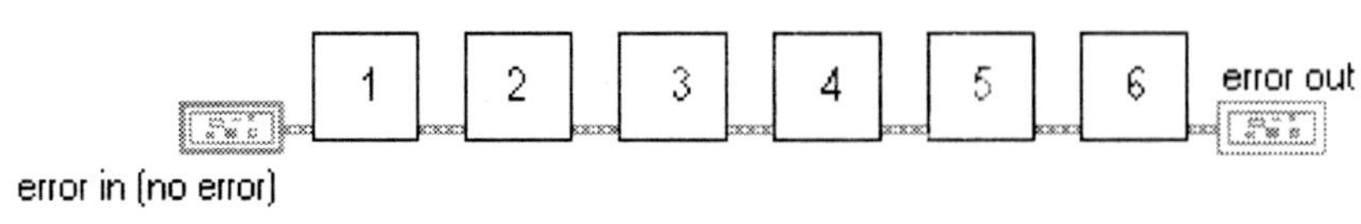

error in (no error)

Figure 6.48
Pass through error handling.

parallel and a reasonably large program, the above method of error handling can lead to unpredictable results.

Because of loose coupling, components have a tendency to be more self-contained than the standard data flow way of writing software. This should be reflected in the method of error handling. Components should be able to handle their own errors as much as possible. Why not have an Error Control Component that can either throw up a dialog or log the error to a file? This component can be placed inside another component or at the end of a sequence, allowing complete flexibility.

There are also areas of your code that are more likely to fail than others. These high-risk areas are usually when your code interfaces with the real world, and if you concentrate your error trapping, in these areas you'll catch a large percent of your nasties.

Areas to look at carefully include:

- Database interaction
- File interaction
- Hardware interaction (especially when controlling hardware)
- Printing
- User input

Lastly, there is severity of error. Is your error an exception or a warning? Is it fatal or just aggravating? Is the closing down of the application an appropriate response to the reported error?

If you have designed your executive as a state machine with each state mapped to a state of the system, why not add error states as well? Are they not valid states for the system to be in?

The following example is an empty test executive that has five test states, an initialize, a reset, and various error states. The order of operation is controlled by the array of Strict Type Def enumerated types. In this example, all the subVIs do is throw an error if the input is set to <Moan>, which can be set from the Local area of the Front Panel. The Front Panel is shown in Figure 6.49.

To change the order of the states, update the array of enumerated types, and to force a failure, set the associated enumerated type to <Moan>, which will activate the required error state. The code is shown in Figure 6.50.

The array of enumerated types is input into the For Loop where it is auto-indexed. The single selected enumerated type is then input into the While Loop shift register and a case is selected. If the case is okay, the "finish" case is called and the next enumerated type in the array is selected. If an error condition is called that case is executed as well.

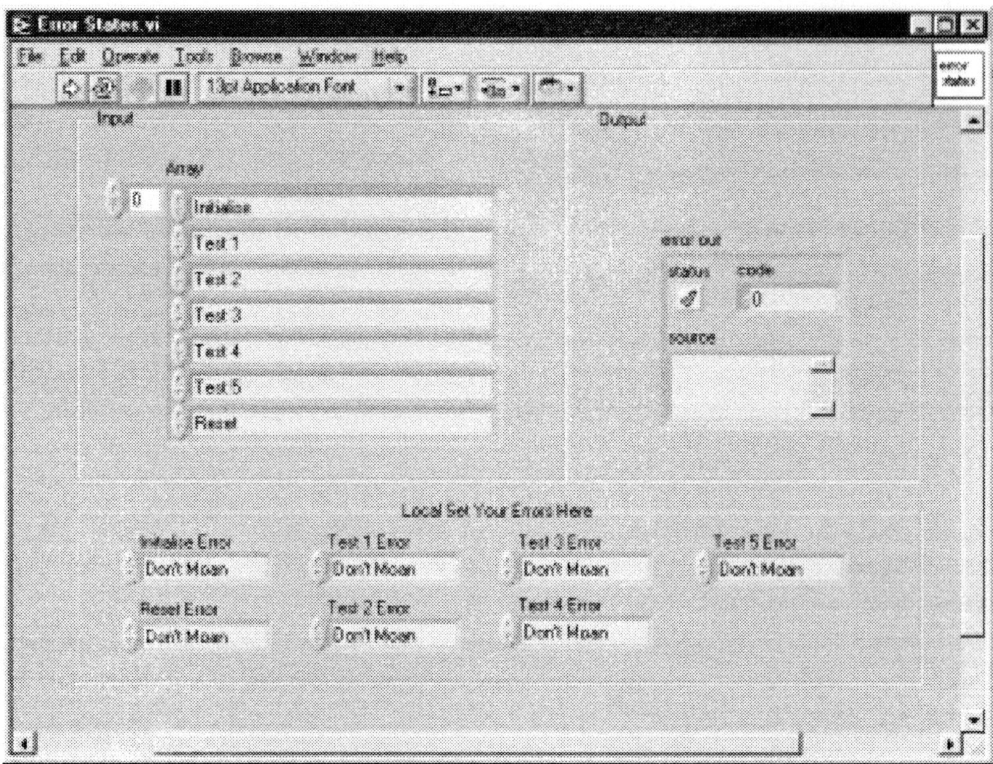

Figure 6.49
Error States Front Panel.

This structure allows the modeling of a systems failure condition to be executed as well as its normal running condition. It also allows errors to be categorized and organized accordingly (warnings, exceptions, and fatal errors). A brief look at the rest of it is shown in Figure 6.51.

6.5 Pre- and Postconditions: Check What Comes In and What Goes Out

This concept is so powerful that we can not stress enough how important it is. The idea, as usual, is as old as the ark, but often little utilized.

Continuing the production line analogy, the production line, in Figure 6.52, just builds and ships. From stores to the dump truck in three simple steps!

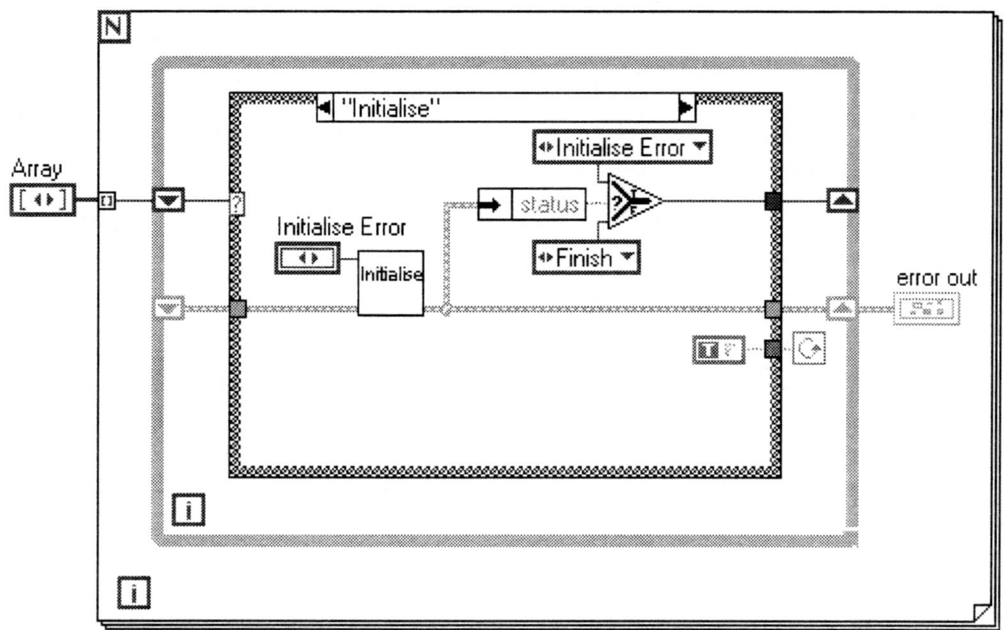

Figure 6.50
Error State diagram.

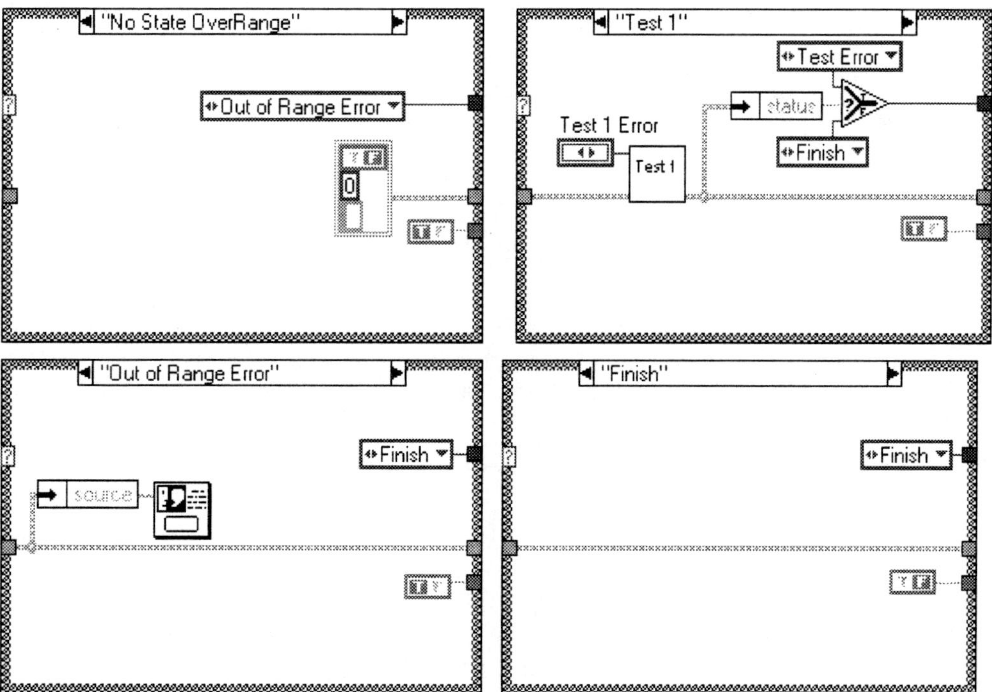

Figure 6.51
Error State cases.

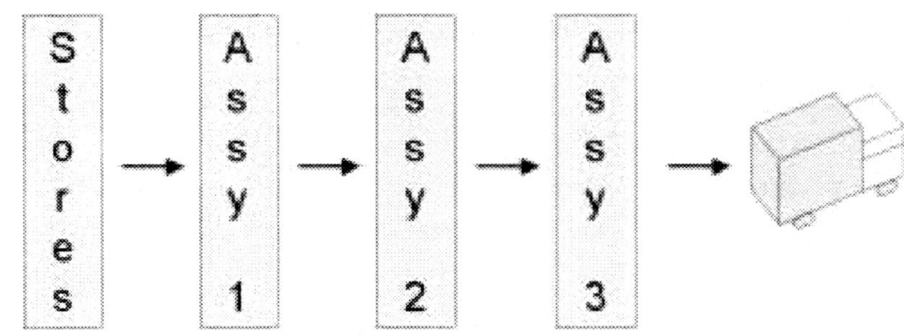

Figure 6.52
Factory No Checks.

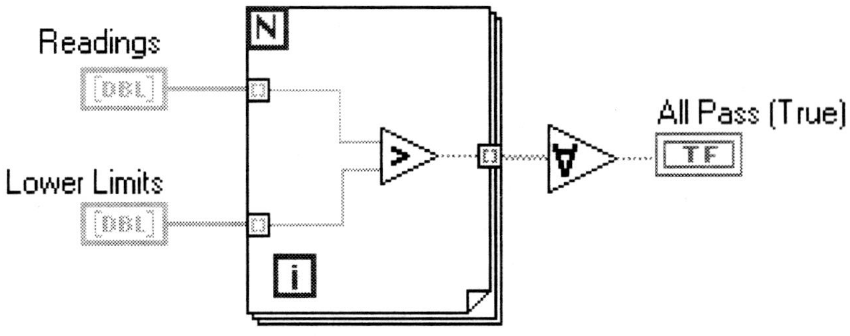

Figure 6.53
Pre- and postconditions example before.

A simple VI example is shown in Figure 6.53.

This all looks pretty straightforward. The readings being passed are compared to the lower limits that are also being passed. Each reading is checked against the corresponding limits as the loop iterates over the incoming readings. The output logical array is all AND-ed and an overall result passed out. So, if all the results are greater than the lower limit the result is True. This could be a check on signal levels, frequency, current, or anything. To add a little importance to the code, the shipment of a product relies on the result being True. The limits are in a database or flat files and are read into memory when the test is kicked off. Simplicity itself!

But, someone makes a mistake. There are 10 readings being passed to this VI, but only seven limits have been set up. What happens? The loop will only iterate seven times, as it is limited to the smallest array (a good thing in itself).

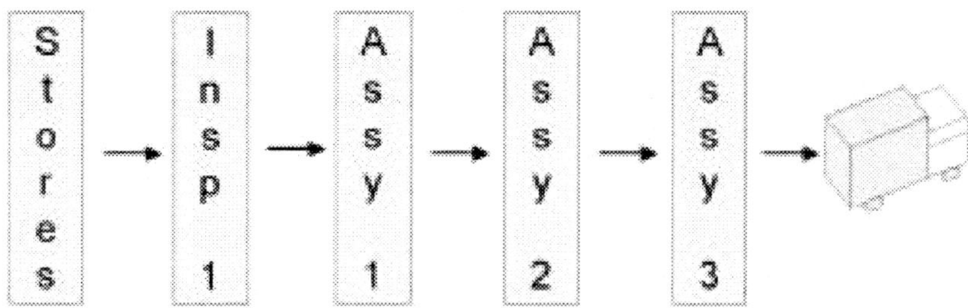

Figure 6.54
Factory precondition.

This means three results do not get tested. If the first seven all pass, the result passed back is true and the product is shipped. This goes undetected, the last three tests start failing, *and we are now in a world of hurt!*

6.5.1 Preconditions

In the factory analogy it would be far better to check the materials coming from stores prior to them being built into scrap. Figure 6.54 demonstrates this. This will pick up any inferior subassemblies and material. This example won't pick up any assembly errors though.

The VI example problems could all be addressed with a simple precondition. In natural language it would be stated thus: The comparison is only valid if both the incoming arrays are the same size.

Sounds straightforward, but the above statement could save a lot of misery when bad production goes out the door.

The code in Figure 6.55 shows one way of implementing the precondition.

Now, the two incoming arrays are checked to make sure they are the same size. If they are not then a warning is given, and the unit is failed (constant in the false case ensures only a false is passed out). That is it, a simple precondition has ensured that any problems or errors farther upstream in your application will not cause a bad unit to be passed.

One common precondition that is well implemented in LabVIEW is the connection requirements shown in Figure 6.56.

Another connector-related precondition we use often is the "No Command —Error", "UnderRange", or "OverRange" enumerated type setting. When a component sees one of these error commands it will bleat loudly. The reason behind this is that if you neglect to set the connection requirements as above,

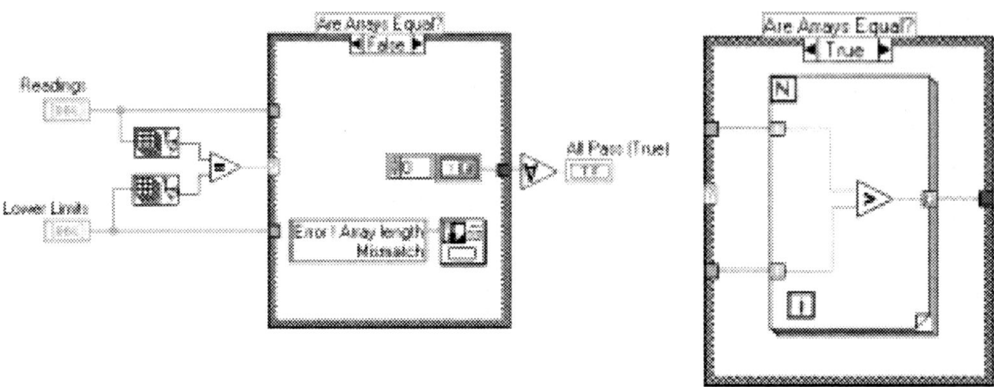

Figure 6.55
VI Precondition.

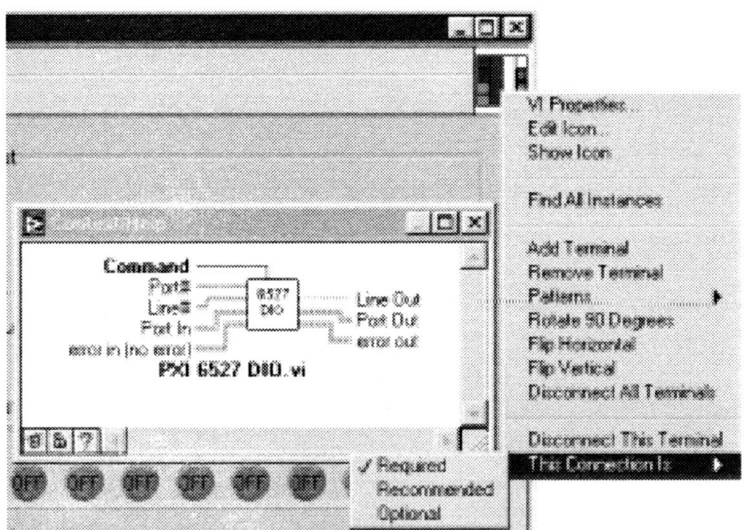

Figure 6.56
Connector Precondition.

and drop the component onto your block diagram without connecting and setting it correctly, you will get unpredictable results.

Another simple and useful precondition technique is to remove the default on case structures. This will break the diagram until all the cases are created.

6.5.2 Postconditions

Postconditions are, well, checking that everything is okay at the end. So, why would we apply this? Again, code that checks code.

Back in the factory, Figure 6.57 shows that we're now checking that the unit is assembled correctly prior to shipping to the customer, but we're not doing anything about scrap and waste during manufacturing.

An example of a postcondition could be closing switches in a switching system. We will assume that this switching system provides the ability to read back the current state of the switches. So, our component commands some switches to be closed. We could then merrily exit this software assuming that the switching system is indeed in the state we wanted. But, maybe it isn't. Our precondition could be enforced by reading back the current state of the switches and comparing them, ensuring that they are the same. If they are not, then we will carry out another operation or simply throw an error. The same can be applied to setting up instruments and then reading back their settings such as environmental settings (set temperature, pressure, etc.).

Read-backs should commonly be employed when doing communications, file transfers, and database transactions.

6.5.3 Conclusion

The hardest problems to find are when the software appears to behave in the manner in which it was intended, and the outputs seem reasonable (in this scenario all production seems to be A-OK). But, by applying pre- and

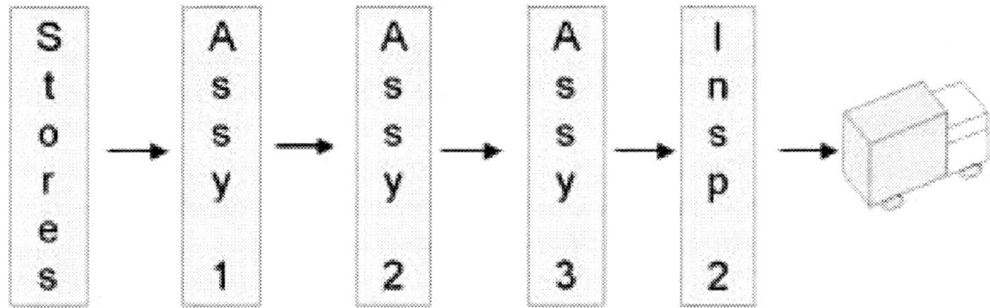

Figure 6.57
Factory postcondition.

postconditions the code is doing a sterling job in continuously monitoring the internal hidden behavior of the system. Essentially, the code is checking itself and that has got to be a good thing.

6.6 Reuse

In the future you'll visit your software supermarket and go shopping for components that you can plug into your pattern, and in a very short time your application will be ready for action. Hmmm. . . . Despite a large amount of effort in this arena it still seems a fair distance away.

Reuse like most good ideas is simple in concept and surprisingly hard work in practice. The benefits do outweigh the effort but it is a long-term strategy.

Why reuse? Essentially, because if you didn't reuse your existing work you'd have to write everything from scratch, which would get very tedious very quickly. There are hidden benefits because reused code has been tested before (if it hasn't been changed). This gives us quality benefits as well as productivity improvements.

What can we salvage?

- Designs
- Source code
- Documentation
- Tests

6.6.1 Opportunistic Reuse

The easiest route to reuse is opportunistic reuse (some gurus call it "informal reuse"). This essentially involves hacking old programs to make new ones, and it is what most of us do to some extent or another.

The biggest obstacle LabVIEW presents is when it picks the wrong file path and you end up killing one of your originals (or even worse changing it without killing it).

If there is consistency of design, opportunistic reuse can be quite effective. Studies have shown that you can expect 40 percent productivity improvement.

But for anything more than a one-man band, leveraging advantage from reuse will take a certain amount of management.

6.6.2 Planned Reuse

This is where a company puts in the effort required to gain real benefits from creating and maintaining a reusable library of components.

Looking at several projects it would be safe to assume that the completely unique aspect of each is in the top-level design and some of the bottom-level detail. The commonality would be far higher than the 40 percent productivity improvement quoted for opportunistic reuse, which implies a certain amount of waste in the process. So how do we get to this cornucopia of high-efficiency state-of-the-art clean room programming? We don't know and its very frustrating. But the solution will come from management rather than from software engineers. Someone has to set up a reuse library and manage it, they will also need to promote its use, police the components that are input, reward or acknowledge the authors, and finally monitor the effectiveness of the policy.

We need to take small steps. Lasting benefits will come from approaching the problem in a careful manner rather than trying to cure all the world's ills at once.

There are things that can be done to improve efficiency and we are going to discuss the tools and techniques that will help.

The speed of developing in LabVIEW already owes a lot to the amount of reusable stuff provided with it. There are also some places to buy VIs, but the market is not well developed yet. The downside of this is the variable quality of the code available. There is also a problem with the consistency of the interface. If you have to learn the code (or tidy and correct it), the benefits of reuse will disappear. One of the advantages of ActiveX components is that they automatically provide and enforce a properties and methods interface. However a commercial marketplace for LabVIEW components will never develop until a common component interface is agreed.

How can LCOD help in moving toward planned reuse. The concept of components is reuse-friendly to start with, and the mind-set when developing cohesive, loosely coupled components is to produce an independent entity that can be reused. By having a self-documenting interface (popping up an enumerated type constant) detailing all of the component's functionality you are again making your component more reusable.

There are also tools provided with LabVIEW that can help, and these are discussed next.

6.6.3 Merge VIs

There is always a risk of damaging the original code when editing existing source code to create new software. If you want to reuse code fragments, there

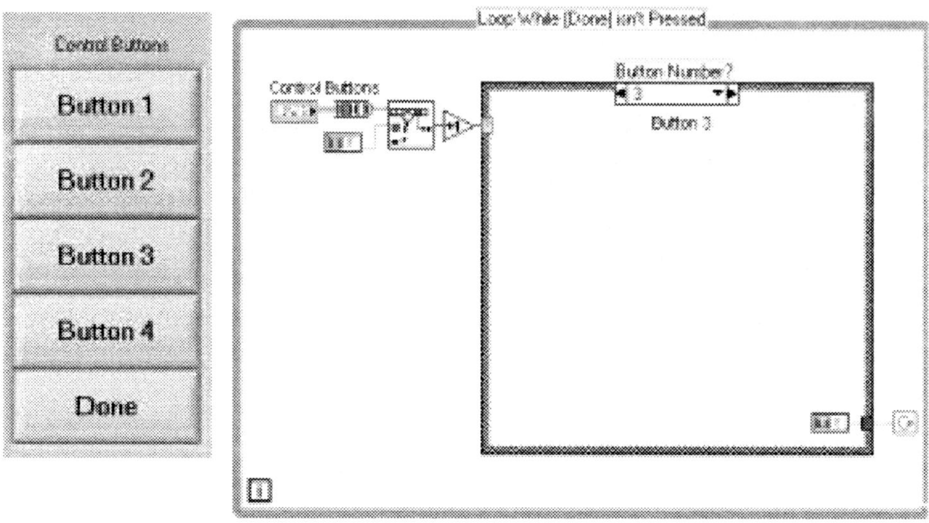

Figure 6.58
Merge VI menu buttons example.

is a reasonably unknown tool that you can employ—Merge VIs. An example will explain all.

First, take a useful bit of code that you find yourself writing over and over again (keep it compact without too many subVIs). Here's a VI for monitoring a button cluster for menu selection as shown in Figure 6.58.

A button press will pass the button number to the case structure. All you will need to do is to fill out the actions in the relevant case. So, for Button 1 you would put your code in case number 1. The button numbers are controlled by their cluster order.

It would save time if this source code were available to drop onto our existing software.

Figure 6.59 shows how to set up a merge VI:

This will allow you to add submenus, VIs, and controls to the Functions and Controls Palettes. Right-clicking on the area where you want to make changes brings up the allowable options.

We will create a new personalized palette by selecting "New Setup . . ." from the drop-down box, as in Figure 6.60

This will give you a default palette to play around with that can be saved under its new name.

Next we want to create a submenu called Merge VIs, which will hold all of our reusable code fragments. Right-click on an empty slot of the palette (you

First select edit palette.

This will throw up the following dialog.

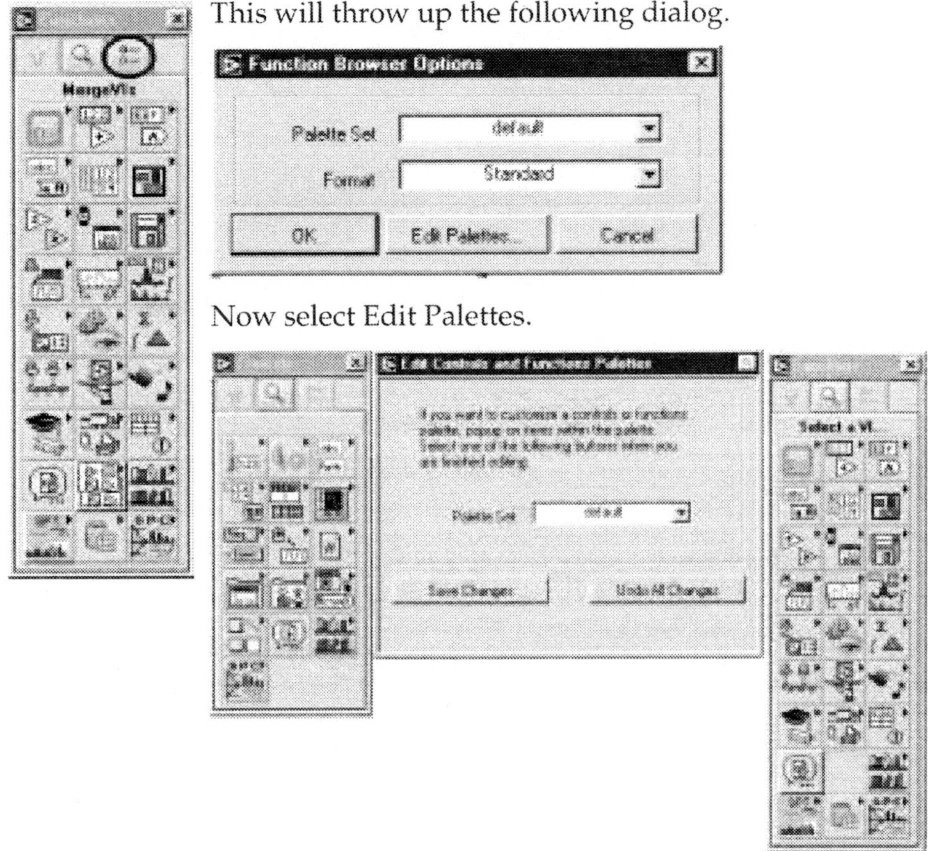

Now select Edit Palettes.

Figure 6.59
Editing Functions Palette.

can insert new rows and columns as well) and select Submenu. . . . Figure 6.61 demonstrates this.

Give it a name "Merge VIs" and change the icon to make it pretty.

You can now open the submenu and perhaps put in another submenu titled "User Interface". In this submenu we'll put our VI. Now insert your VI and select Merge VI, as in Figure 6.62. Save your palette and the job is done.

Now let's see what we've done. Open up a new VI and drag your new VI from the palette as shown in Figure 6.63. Finally, as illustrated by Figure 6.64, you drop it, the code and controls are transferred. Lovely!

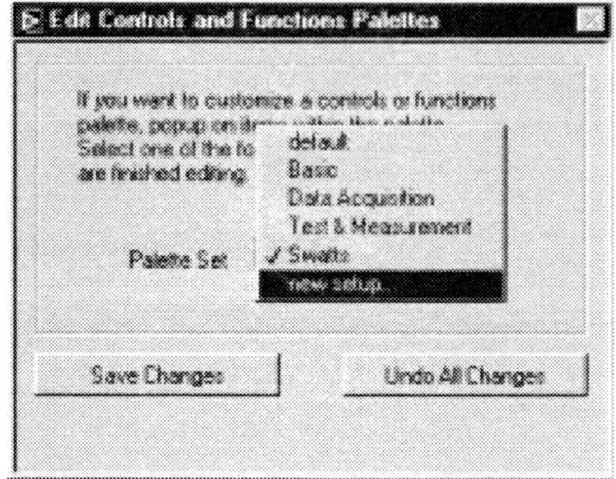

Figure 6.60
New Setup. . . .

Figure 6.61
Insert Submenu. . . .

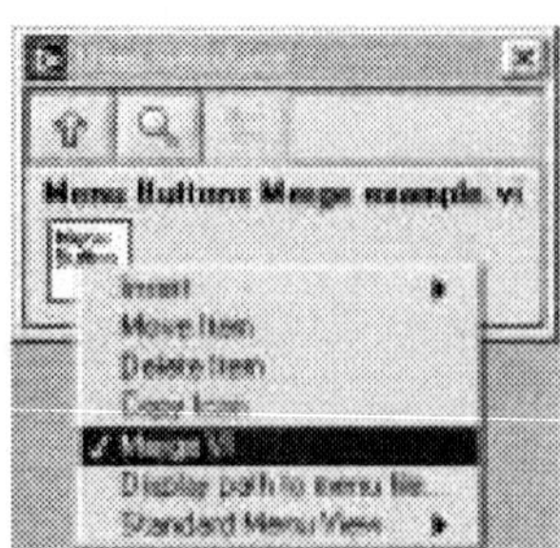

Figure 6.62
Select Merge VI.

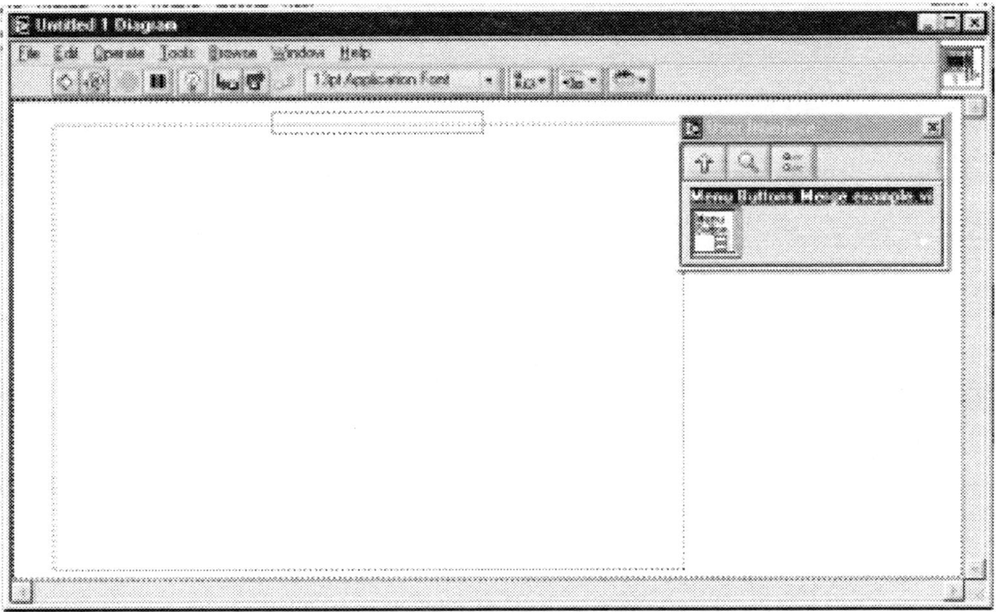

Figure 6.63
Drag Merge VI.

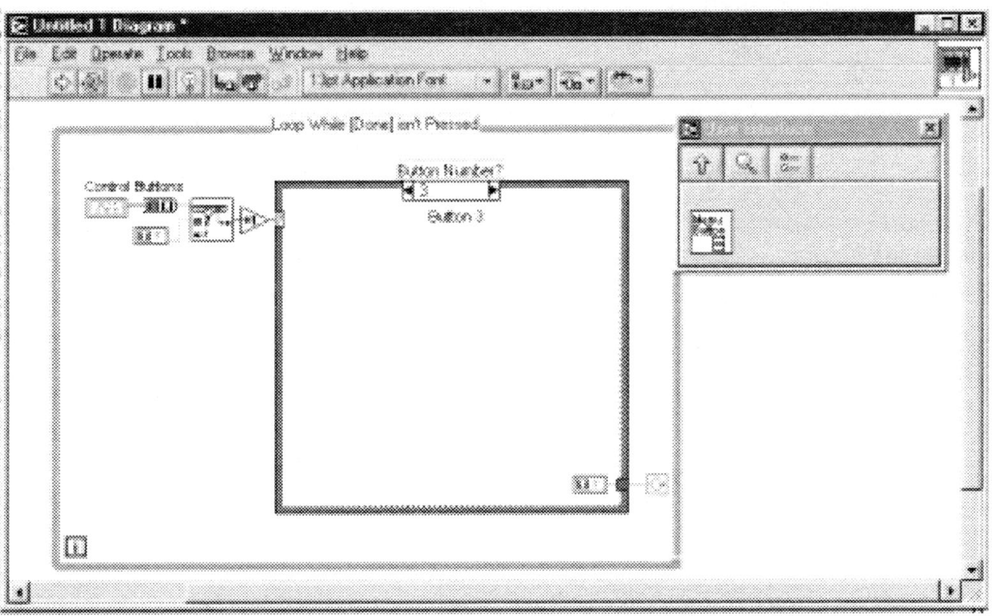

Figure 6.64
Drop the code.

6.6.4 VI Templates

Another way to reuse code and structures is VI templates (.vit).

Creating and editing them is as easy as saving your VIs as VITs. What interested us was more complex VIs that had subVIs and so on. You need to save all of your subVIs as VITs as well (you'll notice that a blue T has appeared on the subVI icons). You can open, edit, and save these Template VIs as normal, but if you select the File>>New..>>Start From Template option and load your template file from here it will create a numbered copy of itself. When you save it, all the subVI templates will moan and require saving as well. The VI templates can be put onto the Function Palette and dragged onto your diagram, but unfortunately they drop as VITs, which isn't a lot of use. If you load them from the Select a VI option they seem to drop properly. So with LabVIEW, if you make template files you cannot stick them in the palette and expect them to act correctly.

From a reuse point of view we found the easiest way of managing reused code was not to share any of it. So each application would be completely isolated from changes in any other. When you want to reuse code, copy the whole directory structure from the original to the new location.

Software Engineering Essentials

7

So far this book has predominantly been about design, and how to bring that design to fruition. Design is a large portion of software engineering, but certainly not all.

Software engineering is such a vast subject that we can only hope to touch on it in this chapter. We left it until the latter part of the book because we wanted to concentrate on the design issues, but before we embark further on our journey we need to discuss the broader issue of software engineering. It is important to understand one size does not fit all. Everything presented on the subject of software engineering is only advice based on the latest information, which means it changes.

What Is Software Engineering?

Engineering is (dictionary definition) the application of scientific methods to the design, building, and use of machines, constructions, and so on.

Software is (dictionary definition) the programs and other operating information used by a computer.

Therefore, software engineering, by (our) definition, is programs and operating data that have been designed by the application of science. Or, in other words, designing software by applying defined methods—the inverse of which is designing software without the application of any methods. The science we've been trying to get across is LCOD. Not tackling the problem with a predefined set of guidelines and techniques, be it LCOD or any other methodology, means the process is ad hoc, or certain to fail!

So, what else is there apart from design? All systems have to go through the same stages: requirements gathering, design, build, verification, and maintenance. Different companies go about these phases in different ways, so what follows is briefly how we do it. We do it this way because we find it works for us. Essentially, what we discuss in this chapter are those aspects of software engineering that are directly applicable to what we are doing, and that is the project.

What you have to be aware of though are software fairy tales and the opposite, software horror stories. Avoid the fairy tales and learn from the horror stories. These fairy tales and horror stories seem to litter software engineering in equal measures.

Fairy Tales

- The latest software is so easy anyone could develop the system.
- The latest 16GL language is going to make this project a complete success.
- The new Disoriented methodology is going to ensure that all software is failproof.
- The new project management tools from IAMALIAR Inc., will make failed projects a thing of the past.
- The latest Case Tool from HeadCase is going to make software engineers redundant.
- The adoption of the Incapable Immaturity Model will make everything work out just fine.
- The requirements are fixed and will not change.

Horror Stories

This list would be exhaustive. There has already been enough in the press recently to show how bad things really can be. But guess what, they are always directly or indirectly related to the fairy tales. Robert Glass's book *Software*

Runaways more than adequately describes the catastrophic failure that can be attributed to believing in fairy tales. So, don't get eaten by the big bad wolf, and don't believe in fairy tales.

So how do we make sure that our projects don't become horror stories? By the application of software engineering methods, with a good dose of reality. We have already touched on the religious experience some individuals have with respect to certain tools, methodologies, and so on. We believe in using what works.

7.1 The Usual Suspects

Any book or article that discusses the software life cycle always starts with a critique of the classic waterfall life cycle as shown in Figure 7.1.

Life-cycle models are meant to show the progress of a basic project from inception to delivery and beyond. The waterfall was one of the first models

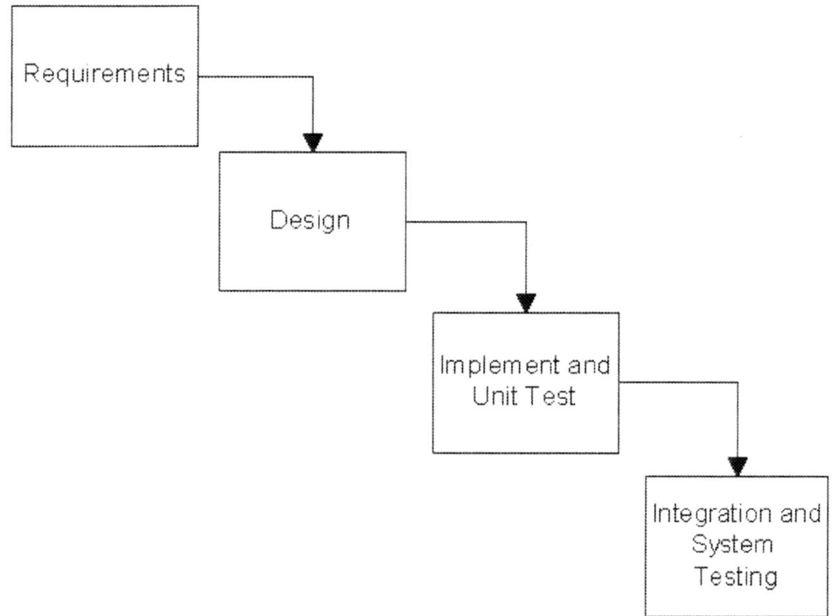

Figure 7.1
Waterfall life cycle.

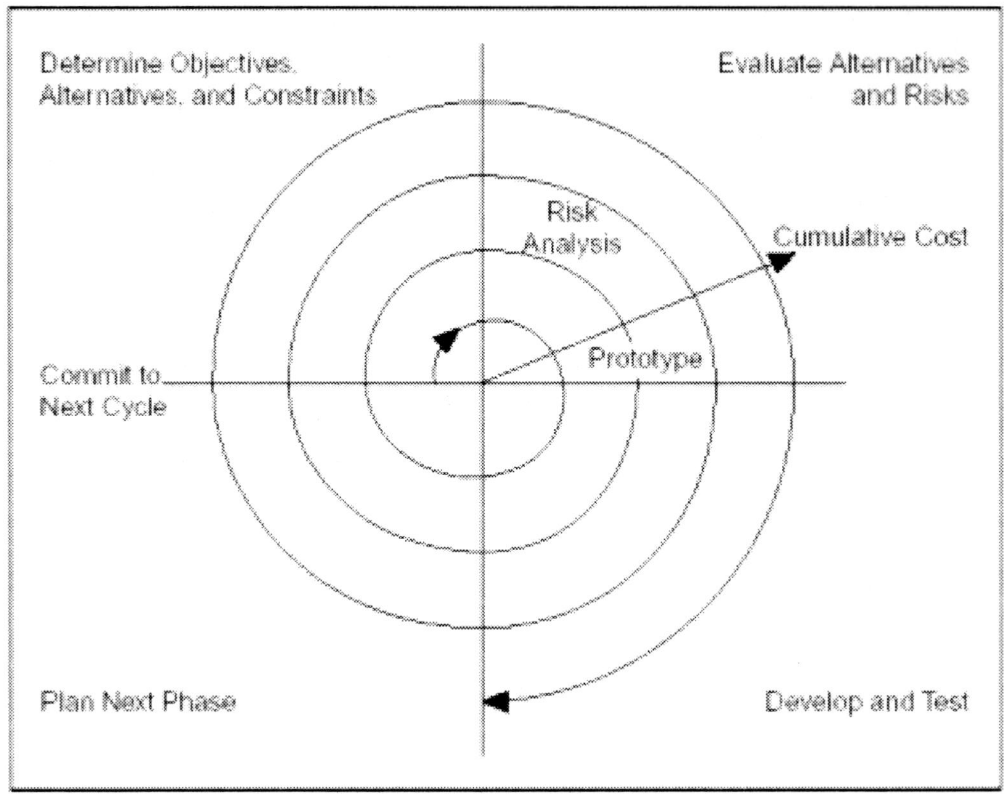

Figure 7.2
Spiral Life cycle.

documented. However, it did not take long for this model to prove unattainable. On the face of it everything looks fine, but it requires each stage to be complete before the next stage can start. Requirements change, design changes, implementation and unit testing can show problems that generate change, and risks are not catered to.

Let us now race ahead and look at that life cycle that is touted as the best alternative, the spiral life cycle that is shown in Figure 7.2.

As you can see, the spiral life cycle is, well, spiral. The main difference is the adoption of risk analysis. If risk of any kind did not exist, then fairy tales would abound. Our problem with this model is that apart from making us dizzy, it is quite complicated. However, along the project's path of progress risks are there to trip us up.

The risks could be:

- Market forces change the requirements.
- It's a new product and the final design is not nailed down.
- The project needs to adopt cutting edge technology that we have not used before.
- New information only comes to light due to results gained in initial testing.

Does this mean we should forget all the traditional specifications writing and just concentrate on risks? No, it means we adapt and use what works for us. We, the authors, can only tell you, the reader, what we use and how it works. What we use is constantly evolving, but the fundamentals are watertight. You must develop the design and construct the software to absorb change. When change or risks rear their ugly heads, you have to evaluate the impact it is going to have, and if necessary, modify your design. You must be flexible.

What is it that enables us to design and construct code in LabVIEW that is able to embrace change? You have already heard this enough I'm sure, but it is cohesion, coupling, and information hiding implemented in a component oriented design. Drum roll . . . smoke . . . lights . . . trumpets . . . LCOD!

We stay away from the big bad wolf due to implementing everything we discussed previously. It is impossible to absorb risk if you have a poorly designed, shabby, rigid set of software. The only way to absorb change is to design your system using the much discussed rules: cohesion, coupling, and information hiding. If VIs A, B, and C are closely related, it is pretty obvious that any change to VI A, as a result of a change in requirements, may well have an effect on all the other VIs. Multiply this by hundreds of VIs and your project will not be flexible.

So, to make our project successful we are going to use LCOD. Perfect, great, we are off to work only on successful projects. Hang on though, that's still not enough.

Software engineering as a subject has evolved immensely through the decades, but the basics still apply. So what life cycle do we use? Our own, although we've borrowed from all the others. We will call it the LCOD Prototype Life Cycle. Actually, we don't have a name for it, I just made that up, but basically it looks like Figure 7.3.

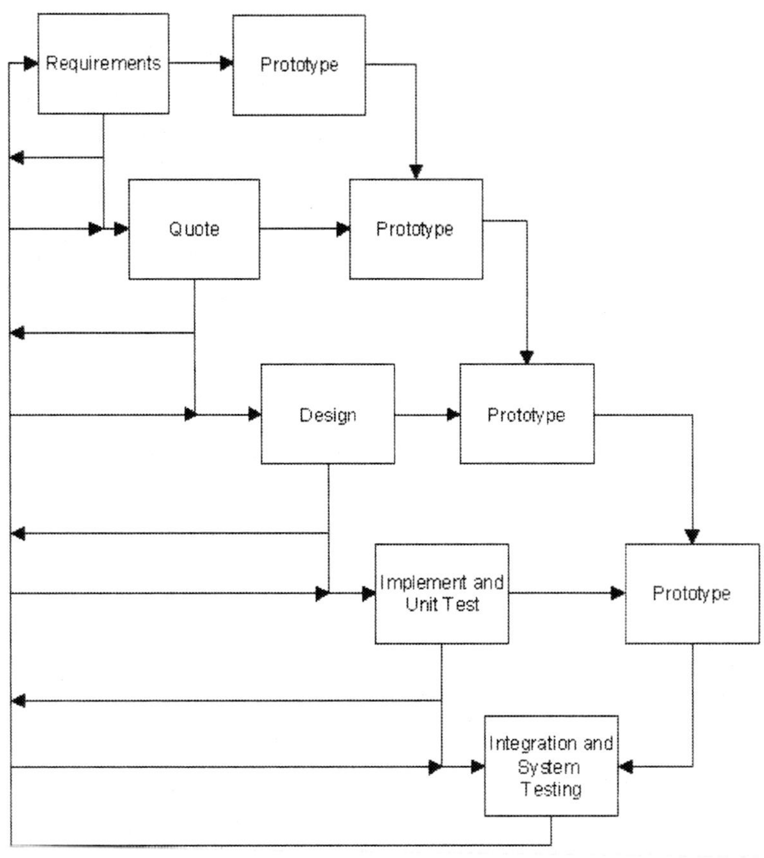

Figure 7.3
Prototyping life cycle.

It looks quite complicated. However, it isn't. At every stage we build or reuse prototypes. Each stage can go on to the next stage or back to previous stages. What we want to end up with are prototypes that have been proved and validated, the sum total of which will make up our final application. Everything pivots around prototyping. LabVIEW as a language is perfect for this, and nothing will build more confidence in the system and your customers than seeing working prototypes that prove the goal can be reached. These will provide convenient milestones. Should you use this model? Maybe, but the main thing to grasp is to keep prototyping and refining.

Do we still write all those reports? Yes, most of them, but they are always a work in progress. That doesn't mean that they are never finished, but that they document both what was understood at the start of a project and what was learned as the project progresses.

The process of continually documenting the project gives everyone the clarity that is needed to ensure that every person knows what stage he or she is at. The collective set of documentation provides the map to ensure that everyone arrives at the final destination. If everyone has a different map, or even worse no map at all, how will everyone know where the end is, never mind how to get there!

All the latest available information states that you should plan with changing requirements in mind. Requirements will evolve and we need to adapt to this fact.

So, armed with LCOD, how do we ensure that a project is successful? The basic starting point is project documentation, which most people hate writing or reading, but they should be embraced. Does that mean we need to write *War and Peace?* Hopefully not. Try and write the minimum that is required, not the maximum. Next we'll discuss what documents we use.

7.2   Requirements Document

A detailed description of what the customer wants, written in his or her own words, is generally called the Requirements Document or Specification. We can't emphasize enough how important this document is, and without it you'll not know what the project needs to achieve. Your project is finished either when you've met the customer's requirements or when your customer pulls the plug!

We can use an easy analogy here. Imagine you commission a builder to build you a house. He has built plenty of houses in the past, so all you need to do is give some loose requirements that you want the house to meet. You tell him or her you want a four-bedroom house with a garage. Off he or she goes and six months later he or she calls you to tell you that the house is ready. You arrive to view your new house, and what a shock. You wanted a two-story house; it's a single story. You wanted a double garage; the builder has built a single garage. You expected a full bathroom; the builder has put in a single shower room. This may seem a bit far-fetched, but just look at how many software projects are delivered, only then to find out that it is not what the customer wanted. The requirements document should mean everyone is aware of what is to be delivered.

A very high percentage of projects fail because of poor requirements gathering, and you will be put under pressure to accept loose requirements. Don't do it! Loose requirements are a big pain and can kill a project and your reputation. Finishing a project with poor requirements is a bit like chasing the end of the rainbow.

And here lies another fairy tale. The customer must write the requirements specification in its entirety before the project starts. Once written the requirements are not allowed to change. Therefore, when you start the project, the customer will have already written a full and detailed requirement specification for you. WRONG! What you will have to do is help gather the requirements. A customer cannot ask for requirements that he or she is not aware of, especially if there are related issues to some of the requirements the customer is asking for. Note the following statement:

"I know you believe you understand what you think I said, but I'm not sure you realize that what you heard is not what I meant."

—*A Customer*

We stated previously that the requirements would already be written. Where there is communication between parties in the software creation process there is scope for misinterpretation and confusion. Your program may be the most elegant, robust, and clever piece of software in existence, but if it isn't what the customer wants it's a waste of time and money.

Requirements gathering is the process of extracting requirements from your customer, and it is also a good initial opportunity for you to impress your customer and to start managing the customer's expectations. Even if the customer already has a comprehensive requirements document, don't believe it. Study it, probe, ask lots of questions, and gather the requirements that aren't mentioned.

There are plenty of studies that prove that poor understanding of a project leads to disaster, and that mistakes made at this stage are the most difficult and expensive to rectify.

The customer needs to have done at least some work up front; it would be ideal if the customer knew exactly what he or she wanted in great detail and had the ability to communicate this to you with no ambiguity. This is RARE! Conversely, your customer expects you to be able to translate the woolliest requirements into a concrete and precise interpretation (that's what he or she is paying for after all). Again, this happens often. The other problem is that customer and developer often speak a different language. These problems are reduced by LabVIEW but not eliminated.

The customer expects the following of the developer:

- Understands what he or she has been told
- Communicates his or her design ideas
- Communicates any problems
- Keeps to schedule
- Keeps promises
- Asks the right questions
- Leads the project

The developer expects the following of the customer:

- Knows what he or she wants
- Communicates his or her requirements
- Is aware of any pitfalls
- Is consistent in stating views

Here are some important points to establish before attacking the problem.

Who is the Real Customer?

This is important because quite often you can be doing work for a customer representative who you think is the design authority, only to have his or her decisions overridden by the manager or another project engineer. If it's possible, get the design authority formalized; it will save grief and finger-pointing later on. Sometimes the manager who has final approval isn't available until right at the end of the project. This is a recipe for a large amount of aggravation.

What Constraints Are There?

Constraints are often overlooked, but they are important and could again create all sorts of problems. Constraints are customer-enforced requirements such as operating system, language, speed, and so on.

How Do You Manage Expectations?

There are numerous myths surrounding software, which can lead to customers expecting more than they will receive. A disappointed customer won't be back! It's a good idea to offer up a few truths beforehand.

- Software isn't easy.
- Design up front is far less expensive than problem solving at the end.

So you've asked the right questions, and your customers are aware of potential pitfalls in the software process. Let's go gather requirements.

Before we start, a couple of quick points:

- Gathering requirements well is not easy!
- Any effort put in here is worth it.

Requirements gathering is all about problem evaluation; it is not about solution synthesis. You need to separate functionality from implementation, concentrating on the WHAT not the HOW.

Another good question is WHY. Question the requirements themselves. It's amazing how requirements can disappear into thin air when the person requesting the feature has to justify the reasoning behind it.

What is the Big Picture?

A basic approach to gathering requirements is that if you understand what the customer wants the software to achieve, you may be able to influence the design process rather than being influenced by the design process. You will also be in a stronger position when the customer changes his or her requirements.

You need to anticipate and identify these changes as much as possible. If you understand what the customer wants your software for, you'll be in a good position to comment intelligently and plan appropriately.

Another major aspect in gathering requirements is sorting out the MUSTS from the WANTS. Often projects fall way behind and costs skyrocket due to the bells and whistles that your average "techno sheep" like to munch on. "Techno sheep" was a term learned from a project manager; he used it to describe the difficulty in getting his programmers to stick to core problems on the project when all they wanted to do was wander around munching on the interesting bits. Identifying and satisfying the requirements that are pivotal to the problem domain are often key to perceived success.

This next point is probably the most difficult to manage. Don't always believe the customer's requirements. Sometimes the customer doesn't have a complete view of the problem and, therefore, may need to be led gently in the right direction. The only other way to manage this is to put flexibility in the system, which will allow for late requirements changes.

Identifying requirements that haven't been asked for is also extremely powerful. A customer once informed us that they wanted all the results in flat files; they would sort out the data later. Their decision was based on lack of knowledge of databases. They added the requirement when the pitfalls of not using a database and the power they would gain from using one were pointed out.

Another example is to make a system totally self-sufficient with respect to the network. You may have experienced systems that die if the network fails because they are using networked servers for information or storage. If the network dies the system stops, which could mean entire production lines coming to a halt. However, if we make the systems independent of the network this will never be the case. We can make them intelligent enough to recognize that the network has failed, so they can default to their own hard drives. This has huge payback, but we have never known a customer to ask for it, even though we have always persuaded them to do so.

It's hard for a customer to request something that the customer does not know exists. It's part of our job to offer requirements for consideration. Another thing to bear in mind is the fact that the only constant in software engineering is change.

If your project is a test system or it has hardware of some sort you're in luck. A test specification can act as a blueprint of sorts. Hardware usually has easily definable states and attributes, which again can be an advantage (keep an eye out for constraints though, and ask WHY).

Using LCOD has resulted in the identification of patterns in software; these patterns can be reused and can help in the formulation of questions to ask. They also allow prior knowledge from old projects to be applied to new projects.

For example, there are questions regarding the User Interface that will need discussing such as Data Handling, Reporting, Hardware, and Tests. All of these can be brought to the table and specific questions asked. You can apply a checklist to ensure that all likely areas have been covered. Do sketches of screens and reports. Identify the different actions and attributes required for the User Interface. Similarly, list all the actions and tests for the Hardware and Tests components.

In an ideal world enough information should come from these meetings to allow an accurate evaluation of system complexity, which will enable you to provide a decent quote. STOP!

We advise the provision of a prototype that allows the customer and users to get a hands-on feel of the system. You would be amazed at the amount of information that comes from doing this. LabVIEW and LCOD give you a huge advantage. You can rapidly provide a prototype that emulates the customer's requirements, and doing this doesn't compromise the software design. You can then sit down with your customer and modify it to his or her needs, usually on the spot (this generally impresses the customer a lot!). Finally, you can take this prototype and build your software from it. This is a technique that we employ for all but the simplest projects.

Another useful requirements gathering aid is writing the Instruction Manual and getting that approved first. We must confess that we have never used this, but it seems a logical idea and may be applicable to more complex systems.

As we have said before, requirements will change to a greater or lesser extent no matter how much time you have spent nailing them down. Be prepared to accept changes as the project moves on, and make sure the requirements document is kept up to date. This document could save you if it is pointed out that a requested feature is not in the software, because if it's not in the requirements then the customer has no argument. Of course, this also means it could be used to beat you over the head if a feature is not in the software but is in the requirements.

Do not accept changes without looking at the impact on your timescales. If the deadline cannot move then something has to give, like other features. The worst thing you can do is not have the requirements written in a formal document and all concerned agreed on them, even if they are changing (which they certainly will). If they are not written down then the requirements are completely fluid, and even worse, they are whatever the customer decides that day, hour, or minute.

7.3   Quote/Project Validation

You have the perfect Requirements Document, now you need to tell the customer how much it's going to cost. How easy this will be can be gauged by how many projects go over budget, sometimes by hundreds of percent. Before you can quote or estimate a job you will need to have done a certain amount of design. There is an uncomfortable balance of either completely designing a system to give a top-notch quote, or not spending enough time on the design and giving an inaccurate quote. Bearing in mind that a lot of jobs run over expectations, the latter can be very painful. However, patterns and past experience can play a huge part in making your estimate more accurate. The absolute worst mistake you can make is not understanding the requirements as completely as possible. It's probably the surest recipe for disaster.

One of the best pieces of advice you can give to a customer is not to go ahead with a project at all. You are being asked to do the project, so surely you should assess the likelihood of success. To go back to our building analogy, you would be sorely disappointed if after the house had been built, you could not live in it because the ground it is built on is the wrong type and the whole structure is sinking. You would have expected someone to check first. Advising against going ahead may go against the grain, but if the project is not feasible for whatever reason (you may have already seen a project like this fail), then it should be reassessed. For example, there often needs to be specific criteria for the project to get off the ground, which could be payback in a production environment. The customer has calculated that by employing the latest Digital Signal Processing (DSP) technology he or she can cut the test time, and the entire new project will pay for itself in one year. This may be impossible, and, hence, the project (as defined) should not go ahead.

The initial quote needs to be as accurate as possible, which is extremely hard when you are predicting the future. Careful consideration of the customer's requirements, compared with your previous experience, should play a huge part in the estimating effort. Prototypes can help immensely, especially with timing or performance issues (the analogy of tracer bullets works well here). If you are not sure that some feature can be implemented, then try it out first. There is, of course, a cost to you for doing this if the customer is not willing to pay for a trial, but that cost could be minimal compared with giving bad advice and being saddled with a failed project. It's far better to tell the customer at the project inception that he or she will not get the payback they expect, than to get that sickening feeling at the end of the project when realization sets in. Do whatever is feasible to ensure your estimate is as accurate as possible in every facet. Alternatively, just pray.

7.4 Target Specification

The target specification documents how the system will be produced, and what technology will be employed. It's usually quite short, but no less important. This should detail the software development platform (LabVIEW of course), computer vendor and specification, measurement devices, connectors, and so on. To use our previous analogy, you go to view your new house and it has been constructed using wood; you assumed it would be brick, but it has not been specified.

If your customer expected LabVIEW version 99 running on the latest Intel, but you deliver the system running on an old 486 DX that you found at the back of the cupboard, written using an ancient copy of LabVIEW version 3, there will definitely be tears. Your customer is not happy because he or she had expectations that were not met.

We also try and put all main user interfaces into this specification. It will force the customer to get involved with screen layout at an early stage and raise new issues about events and functionality. This should be one of your major prototypes. Build the screens and simulate their behavior, review with the customer, and when acceptable, document it here.

We find it useful to put all the timescales and milestones in the Target Specification. As its name indicates, this is what we are aiming for, so it makes sense for them to go in here.

In summary, the Target Specification should detail all expectations and how they will be addressed, and eliminate any assumptions about the system to be delivered or the project as a whole.

Assumptions can kill projects and should carry a government health warning!

7.5   Test Plan

How do you know when you have finished? Sounds like a stupid question, but it is when there is no more software to write of course! Wrong. You are finished when you meet the test plan. The test plan should detail exactly what the system should do, what the interactions are, the expected outputs, and so on. It should exercise the system in normal scenarios as well as abnormal, so there are no hidden surprises. Most importantly, it should prove that all the requirements are met. If missing or poor requirements are the primary reason for projects failing, then lack of a test plan is second. We have been involved in projects that have never ended because no one knew where the end was.

Another pitfall is writing the test plan at the end. Come on! Why an earth would you write it before the software has been written? The answer is simple, all these documents need to be agreed upon, which means they give a common discussion point for all concerned. You need to get everyone to buy into these documents, all expectations should be thrashed out, and then all will know what is expected.

Table 7.1 *Button and lamp tests*

Action	Result	Pass/Fail
Press button X	Lamp X lights	Pass
Etc. . . .		

Having the test plan at hand will aid the software construction process immensely. Whenever the developers raise questions about how a certain aspect of the system should perform, consult the test plan. It will aid the entire process. It will also enable unit testing to be clearly applied, since it will not only give a clear insight into the overall goals of the system but also gauge the performance of your components as they are built.

All of our test plans use the same layout, a table with three columns as shown in Table 7.1.

This makes all tests simple to evaluate and gives a clear indication of completion and to what criteria.

How did you ever live without it?

7.6 Software Architecture Document

This document will describe how the software will be built. Here you'll include feasibility examples, database designs, architecture designs, test specifications, file examples, prototypes, and more. All the effort here will be to achieve as full a description of the system as possible, without actually writing it. These are our plans for construction.

This activity can take a long time, depending on how much content you put into it. This is one of those areas where trial and error will come into play. We have seen projects where a whole rain forest has been used to print the architecture document of the system, only for it to be filed away for years in a drawer. Other times it can be just a sketch on a cigarette package, telling anyone looking at it little about the how the system was designed.

For us, the main purpose of this document is to detail the design decisions we made. Going through the process of designing the system should raise more problems and issues that need addressing, giving you a chance to make decisions or investigate alternatives. It should also give you a clear plan for construction, detailing the individual components, what their purpose is, and how they communicate. This is also the place to document issues that can prove to be rather gray, like behavior. For instance, we have covered state machines, but sometimes their states can be complex. To decide how they should interact needs work, and that work should be done and documented here.

We also do LCOD design in this document. We utilize all we have learned, and use it to design the system. We derive the components that we will use to build the system, including everything we discussed in Chapter 4: top-down, bottom-up, and OOD. Everything that helps us take the problem and dissect it into a solution.

So we would expect to see:

- Flowcharts
- State transition diagrams
- Schematics
- Relational database design
- Files and a definition of their layouts
- Components and a definition of their interactions
- Pictures of User Interfaces

7.7 Software Construction — Build

Roll up your sleeves and get down to what we all like doing. In the construction stage is all the work needed to convert our designs into a customer solution. Essentially, it will consist of developing, testing, documenting, and integrating your components. The previous chapters have covered this in depth.

From our specifications we know what the requirements are, we know the target, and we know how the system is to function from our test plan. Finally, we know how to build the software from our architecture document. Easy! No, not easy.

Unless this is a repeat project you will be breaking new ground, which is unexplored territory. However, patterns, LCOD, and prototypes will reduce the risks.

All things being equal, this part of the project should eventually become one of the short phases. By using our experience, reuse, patterns, and so on, the construction stage should become shorter the more proficient we become.

LCOD has helped us along the way. We build the components, test them, and demonstrate them to the customer as we go along. Why? It helps us get paid, because we can demonstrate that we have met the identified milestones. The customer can see progress, which gives him or her a warm feeling.

7.8 Test—Customer Acceptance

We've finally made it. We test the system against the test plan. The test plan was written way back at the start, so there are no surprises. Even better, we've been testing the system against the test plan as we went along, so confidence is high. You and the customer run the test plan and everything is fine. What a wonderful project!

In reality, all these documents and stages will change as the project progresses. A natural consequence of a project is that you learn more about what you are trying to achieve the closer you get to the end.

7.9 Pictures Tell a Thousand Words

One way of keeping a check on how much you write is to draw pictures. There is no more common form of language than pictures. Which do you use? Pretty much whatever is going to get the message across the best. Unified Modeling Language (UML) does not convey a menu hierarchy as well as a good old-fashioned flowchart. However, a flowchart cannot convey the complexities of a detailed overall design. Use what works. The most common diagrams we use are data flow diagrams (DFDs), flowcharts, and STDs.

7.9.1 Diagrams—Data Flow Diagrams (DFDs)

Steve Watts writes—"From the mid-1980s to the early 1990s I only knew of one graphical notation to use with software, the flowchart. I found that flowcharts were fine for certain aspects of design, but they were too detailed for more complex systems. They were also inefficient for mapping systems with many conditions and their branches. It seemed that this was a good enough reason to do no design at all."

In the late 1960s Edsger Dijkstra and associates launched an influential attack on unstructured programming, stating that unconditional branching (GOTOs) caused software to become unwieldy, error-prone, and difficult to maintain. They asserted that "Sequence," "Condition," and "Repetition" were the only constructs required when creating software. These three constructs are fundamental to structured programming. The diagrammatic notation for structured programming is called the data flow diagram (DFD) and is very simple. The main symbols for the data flow diagram are shown in Figure 7.4. This simplicity is one of the great advantages of this method.

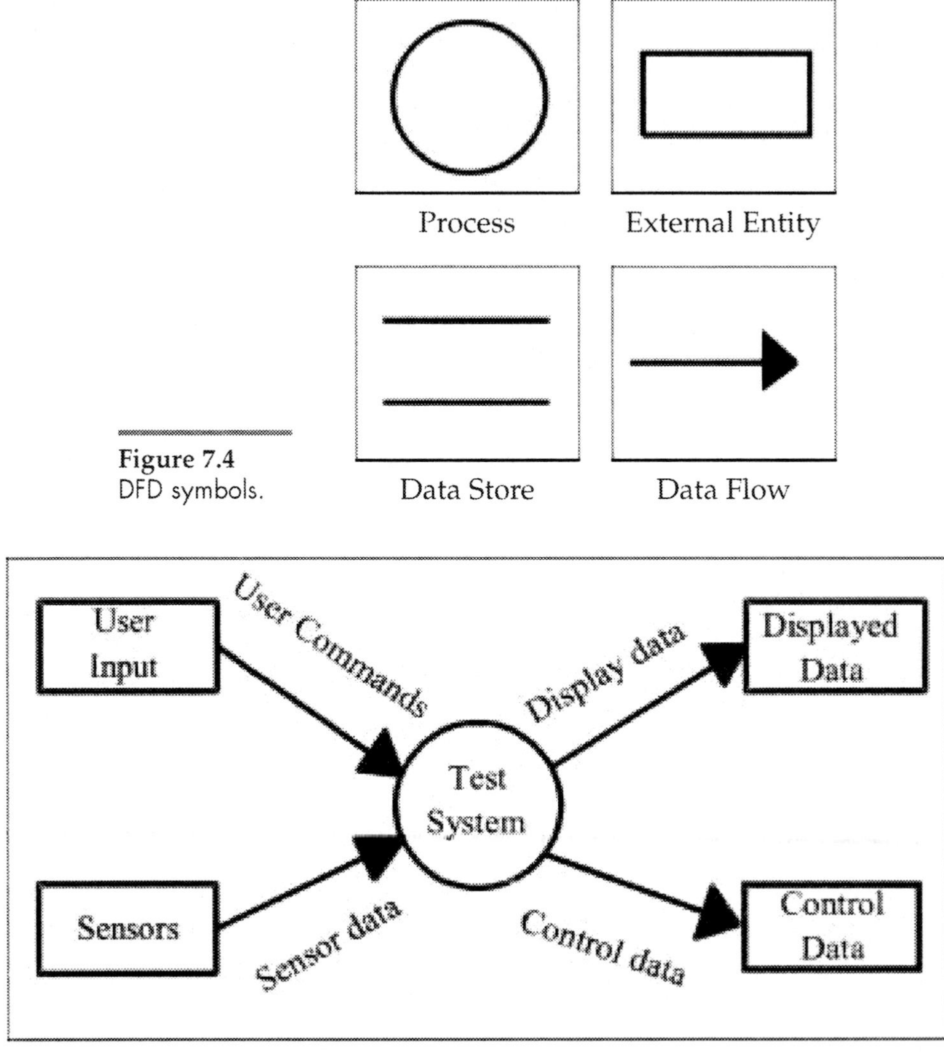

Figure 7.4
DFD symbols.

Figure 7.5
Test system context diagram.

Structured design is about processes, inputs, and outputs. The top-level diagram (context diagram) describes the system in relationship to its inputs and outputs. The external inputs originate from input terminators, and any external outputs from the system end up at output terminators. A typical top-level DFD would look like that shown in Figure 7.5. It's useful to make a list of the data flow contents (e.g., Sensor Data[sensor1mA,sensor2mA,sensor3Vdc]).

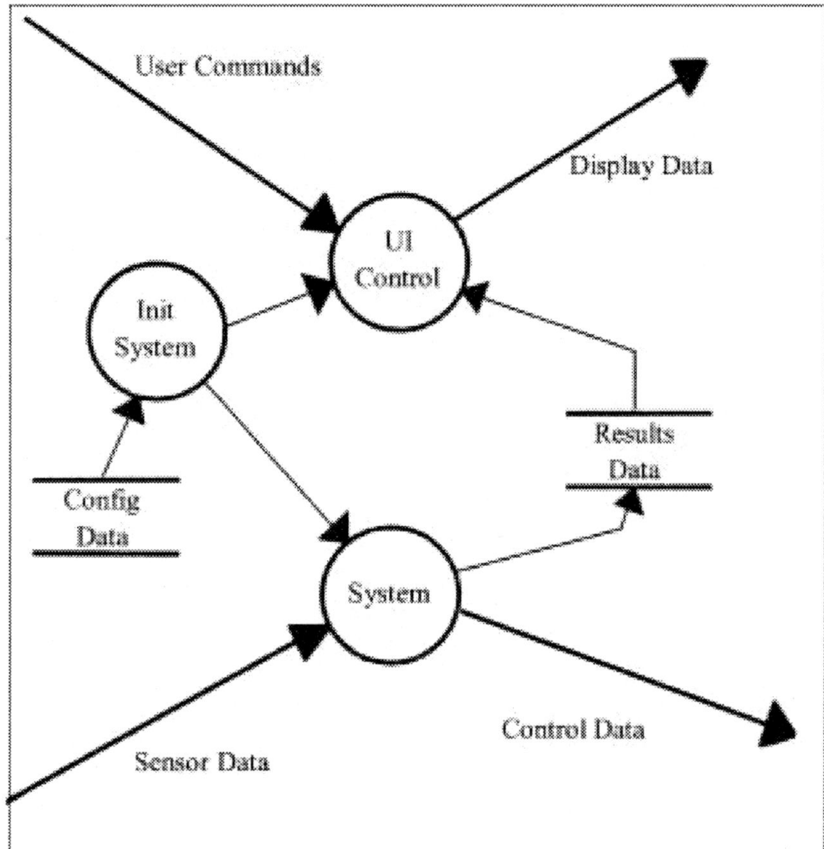

Figure 7.6
Test System DFD.

Another advantage of DFDs is that the diagrams can be decomposed to more detail. Let's look at the next level, "Test System," as shown in Figure 7.6.

All of this is very applicable to LabVIEW, being a data flow language, but you may be wondering why bother? LabVIEW already has a graphical notation. Why introduce another? We find it good discipline to separate the initial design and not to pollute it with any actual programming (plus, I can sit in the garden and do it). This allows us to concentrate on the design and not clutter up our minds with the implementation (syntax). We use it to identify top-level components, or as a communication tool for large, complex multideveloper systems.

7.9.2 State Transition Diagrams

State transition diagrams are very useful for modeling the states of a system and illustrating the actions of a state machine. We mentioned them previously in our discussion of state machines, with the washing machine example.

The notation is simple: a labeled box is used to signify a steady state, labeled arrows are used to signify the transition to another state, and small circles indicate the start and stop states. Looking at a real-world example may help the explanation.

This system was used to control the operation of a complex piece of equipment. For each state certain buttons and conditions need to be monitored and certain ones ignored.

So looking at Figure 7.7 we can see that there is a direct transition from the Start state to the Initialize state. The Initialization Complete transition then moves the system to the Idle state. Here we set the valves and indicators to their idle condition and wait for a press of the "Enable Vacuum" button. A press of the "Enable Vacuum" button causes a transition to the next state that pulls a vacuum on a mold. The valves and indicators are now set for this state, and it sits and waits for conditions to trigger new transitions.

This diagram was invaluable in the design of the system because we could present and discuss it with customers and when agreed map it easily to a state machine structure in LabVIEW.

7.9.3 Homemade Diagrams

Can't find a notation that you like, then make one up. As long as it's not so cryptic that no one else can understand it, then use what works for you. We have a number of homegrown diagrams that we use, one of which we call the Component Interaction Diagram. When we are breaking down our design, we will often show the components that interact to give a better grasp of what components are needed and how they will be coupled together. See the real-life example in Figure 7.8.

This diagram shows all the top-level components for a system we designed. The system was for calibration. Each circle denotes a component, and the connecting lines show their interaction with other components in the system. It looks remarkably like other diagrams, but because it's ours we can dictate the rules. Some might say that this is a waste of time because so many notations already exist, which is a fair point, but always use what works for you.

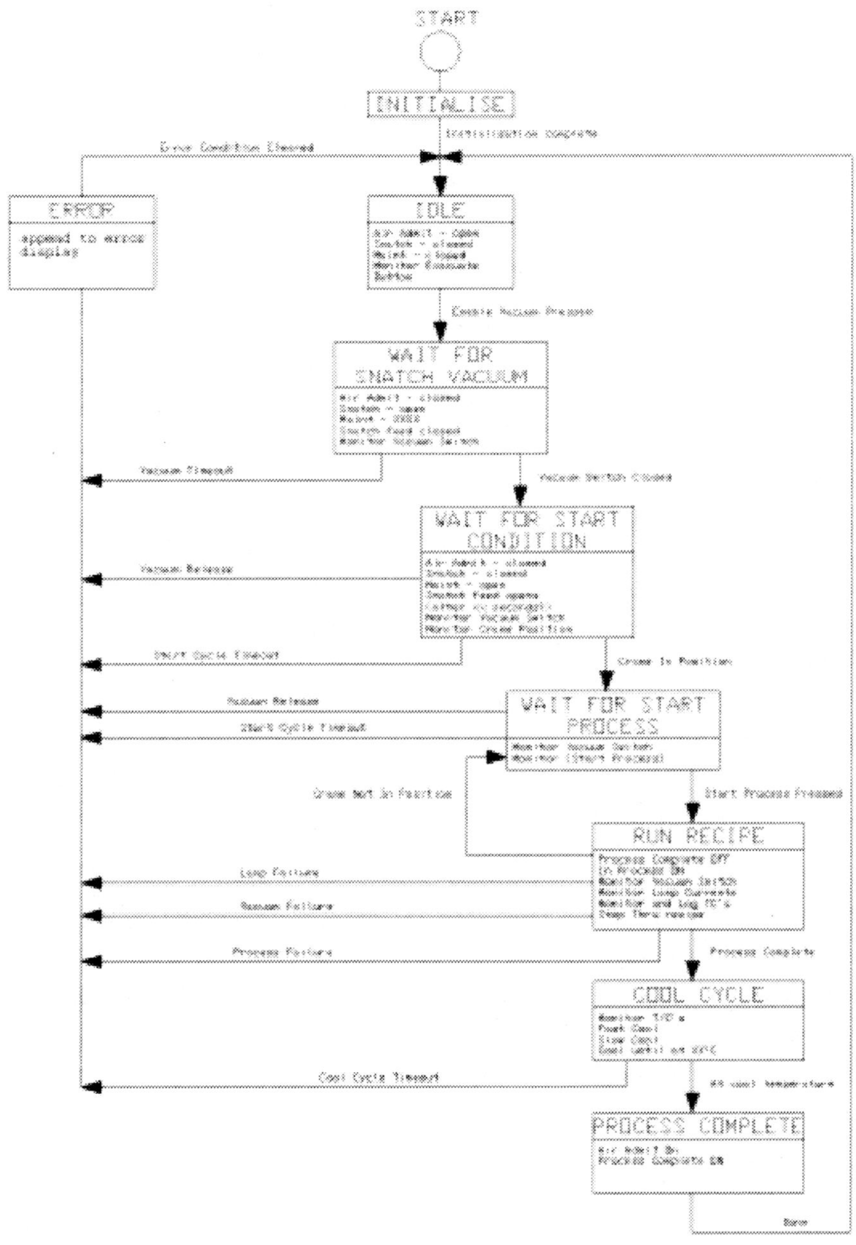

Figure 7.7
State machine example.

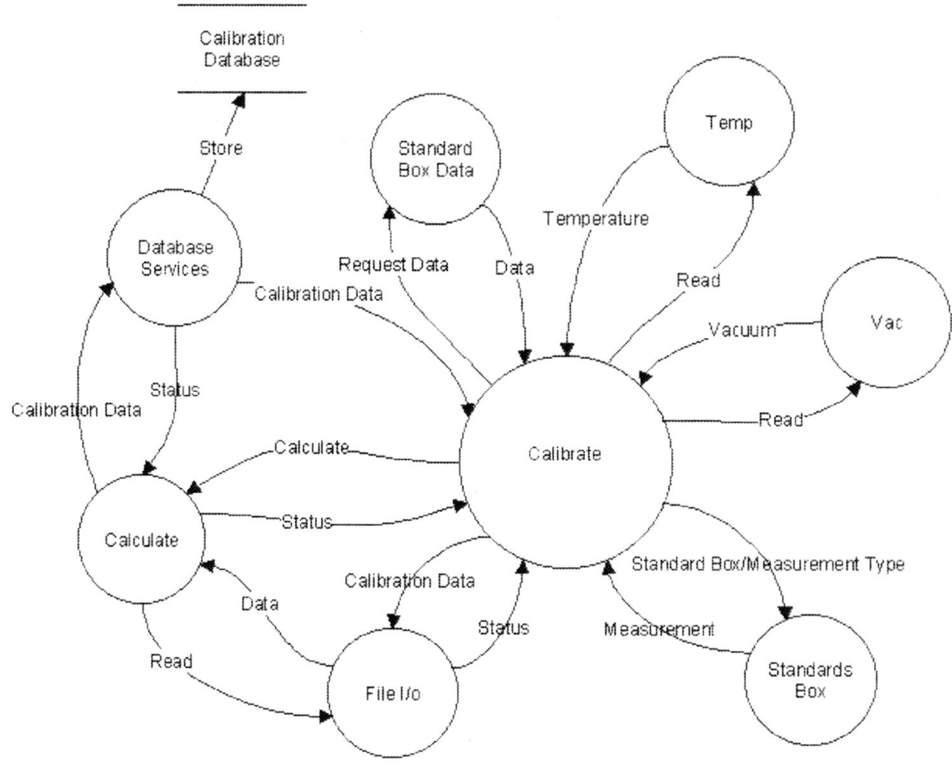

Figure 7.8
Example component interaction diagram.

We try to make our component interaction diagram reusable. For example, to aid estimation. The diagram could also be used in the quotation and then to form part of the architecture document.

7.10 Checklists

Have you ever gone shopping knowing exactly what you want, only to return home without some of the items? The same goes for software projects. Checklists can prove vital to the smooth running of your project. There should be a checklist of items that you can employ and add to as more project experience is gained. In fact, checklists at all stages can help. Everyone often thinks that they know what is required, but placing that information on a checklist will help immensely. Pilots spend years in the cockpit doing their preflight checks, so you would think they know it all by heart. They still use checklists.

Suggested checklists could be:

- Common requirements that appear on all projects
- Common questions that need to be asked
- Hardware questions
- Items to be covered in meetings
- What dictates a VI as being finished

Use checklists as much as possible, they can prove invaluable.

7.11 Code Reviews

The prospect of having someone trawl through your code looking for defects is enough to strike fear into the heart of any developer. Done correctly, many defects can be uncovered at an early stage, thus saving time and money. Done badly, the process can degenerate into a painful exercise, leaving individuals bitter and twisted.

As a brief example, consider the situation where you have a problem in your code that you can't track down, so you ask a colleague to have a look with you. A couple of minutes into explaining the problem and viola, you've spotted it and your colleague hasn't even uttered a word.

So what do we hope to achieve from code reviews? It certainly shouldn't be the stage for the reviewers to impress on the whole world how good they are at the expense of everyone else. What we hope to achieve is to spot problems early, but also ensure that all concerned are following the same style, processes, design, and implementation methodologies. We can collectively call this our coding standards. An often indirect gain from reviews is communication and learning. It more often than not transpires that a new technique or method of doing something comes out at a review, which ordinarily would not have been communicated to everyone else working on the project.

We also feel that it is important that everyone on a team has the chance to become a reviewer. This is good personal development, but also will inhibit any "them and us" feelings since everyone on both sides is participating.

There are several different kinds of code reviews, but walk-throughs, code reading, and inspections cover most. They are all similar, but inspections are the best defined. Without giving a complete discussion of a software inspection, the exercise should include the following:

> Each member of the review team should be given a copy of the software to be reviewed sometime before a formal meeting is held. This will give the reviewer the chance to go through the code and make notes on actual or anticipated defects.

> Each reviewer is fully aware of the design and objectives of the software and the project. If this is not the case, then the reviewers are given enough time to familiarize themselves.

> Each reviewer will have a set of guidelines of things to review, which could be a checklist. The idea is to focus everyone's effort.

> The person who is having his or her code reviewed should have adequate time to review the comments made.

> When the code is ready, a meeting should be held and the comments discussed. It is not necessary to come up with fixes there and then, but to move along at a brisk pace and agree on whether the comments are justified or not. It could be the case that a comment made by a reviewer is an observation that does not constitute any action, once the developer has explained the reasoning behind the code or direction taken.

> Someone should lead the entire effort, especially the meeting. This person is generally called the moderator.

> Someone else should document the problems raised and any further comments raised.

> At the end of the review, a document should be produced as soon as possible that details all the faults and bugs found. This will provide the action list.

> The developer then goes away and addresses the points raised.

> Someone should then confirm that the action list has been completed.

In our opinion, the entire process should be carried out in as lighthearted manner as seems fit. Getting a group of developers to buy into having their code reviewed is hard enough, but then to make it a formal stage for public execution means that everyone involved will resist as much as possible. In the end, this means the whole review system will die and no benefit will be gained.

The published statistics for the payback from inspections are shocking. It has been reported that having inspections can catch 60–90% of defects—now that has got to be a good thing. Not only that, but if you know your code is going to be reviewed, then you are less likely to take shortcuts because as you know they probably will be spotted. The sum total of this is better quality.

To end this short section on inspections, you will find a checklist for inspecting LabVIEW code. LabVIEW is probably far easier to review than text-based languages. Its graphical nature should mean that cryptic- and syntax-like problems cannot lurk hidden within the depths of any code.

LabVIEW Code Inspection Checklist

- All code runs without reported errors.
- Front panels are well laid out, and inputs and outputs clearly defined.
- All user-facing front panels and user interfaces are clear and obvious as to their function.
- Each icon is meaningful.
- Each VI name is meaningful.
- Each component exhibits strong cohesion and is loosely coupled.
- All interfaces to components are clearly defined.
- Each VI has the description section filled in.
- Each diagram for a VI is clearly laid out and labeled.
- The diagram is not so large as to inhibit reading and understanding.
- The developer has reused available code where applicable.
- There are no global variables within the system code, except where formally agreed.
- Preconditions and postconditions are used for all critical VIs.
- All enumerated types contain meaningful entries.

The checklist could go on and on. There could also be project-specific items added or whatever else you feel is going to cut down on the chance of problems arising later. Ensure that problems are fed back in the system.

7.12 The Project Is Dead, Time for a Postmortem

Postmortem sounds very sinister. What this means though is to go back and evaluate the project. See where you made mistakes, made savings, found new information, and the like. Collect it, document it, and feed it back into whatever process you use, refining it as you go along. As we learn over time, we will not make the same mistakes and will gain advantage over our competitors.

7.13 Metrics

What are metrics? They are measurements of your software development. They are also quite subjective. For instance, if you counted the number of VIs in your application and used this to measure complexity, one huge VI would misguide the process. We feel that the best metric you can start out with is from a business perspective, that is, estimated hours to actual hours spent doing a project. If you are way off with this estimate, nothing else is going to help you because you will go out of business. What else can you measure? Quite a lot, as follows:

- Number of VIs (as long as you don't use one large one)
- Number of changes to a VI
- Number of defects
- Amount of maintenance
- Function points
- Amount of reuse
- Anything really, but as always, use what works and throw away what doesn't

It's All About Style

8

8.1 Why Do We Need Standards Anyway?

Why do we need standards? The simple answer is that your source code is how you communicate your design. Any technique you use to improve this communication has to be worthwhile.

Somebody, usually under severe pressure, will pick up your project with maintenance in mind. Will they curse you or kiss you? If you have commonly enforced standards within your team you'll find that life will become easier, sharing will become more natural, maintenance will be easier, and productivity will improve. In essence, pick a style that you and your colleagues are happy with, formalize it, and stick to it.

A good starting point for this information is in the LabVIEW Development Guidelines, which are part of the LabVIEW online documentation.

This section will describe some commonsense standards that we have found helpful, and it will also showcase some of the bad practices and pure laziness that we have come across in our own journey.

Experience the horror of opening a block diagram such as that presented in Figure 8.1.

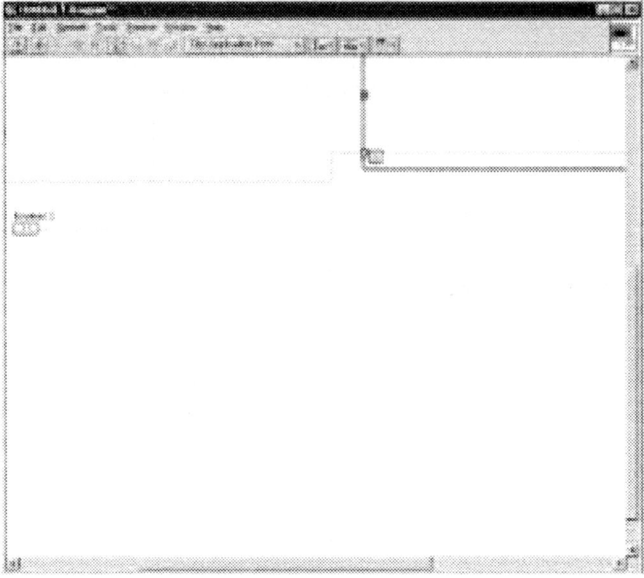

Figure 8.1
Bleaaaaargh!

We enjoy having to walk the mouse the equivalent of 10 miles to find the boundaries of the diagram as much as the next developer, but this is just a pain! It must be more difficult to write programs like this. Diagrams like this communicate a great deal about the perpetrator—that they are an ignorant fool is just one.

Rule 1 then:

WHITE SPACE IS EVIL, DESTROY NEARLY ALL OF IT.

A front panel like the one shown in Figure 8.2 usually accompanies the previous diagram.

Rather than actually spend the 20 seconds tidying this up, let's sprinkle controls and indicators liberally over the front panel. It is hard to believe that there isn't enough time to select the controls and then align them.

At the other end of the spectrum is the diagram shown in Figure 8.3.

One of the advantages of LabVIEW is that the block diagram helps to communicate the design. This diagram fails to convey much because you

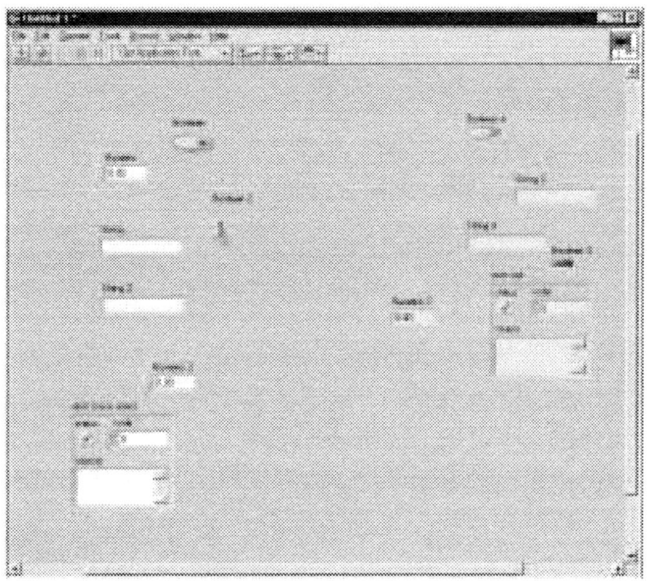

Figure 8.2
Bleaaaaargh! Part 2.

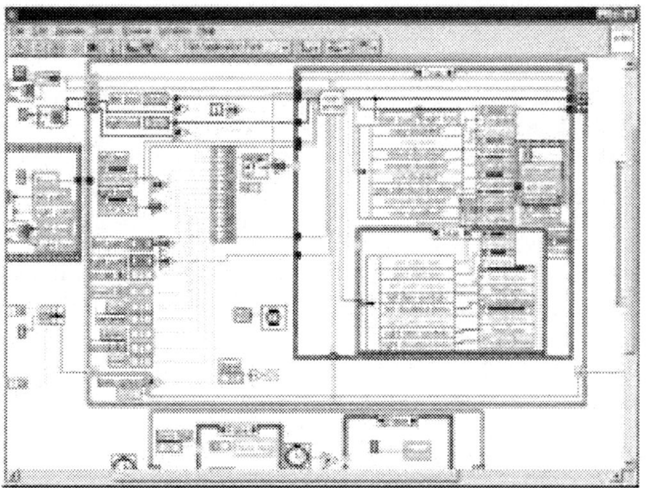

Figure 8.3
I can't see the wood for the trees.

overdose on information. We call it cramming, where you try to stuff as much as possible onto the page.

```
Itseffectsareabitlikethismakingitverydifficulttoseewhatsgoingon.
```

8.2 Block Diagram

8.2.1 General Layout Standards

- Make diagrams as compact as possible while still having a visible flow.
- Don't cram your block diagram.
- Enforce Sequences. Don't leave structures floating. As well as looking bad it can lead to inconsistencies in the program.
- Place any terminals that have no wiring in the first structure of the diagram. They can look like mistakes if left floating.
- Try and keep the overall size of the diagram to one page of your screen. If it's bigger, try and keep 1 page wide or 1 page deep so that you only need to scroll in one direction. A page can be whatever your standard screen resolution is; we try to keep it to 1,280 by 1,024 pixels.

8.2.2 Wiring Standards

- Use left to right data flow. Resist the temptation to stick data into structures from the top or bottom.
- Wires should only overlap if there is no alternative.
- Keep wires as straight as possible.
- Never route wires through an icon to a terminal on the other side of the icon.
- Never route wires underneath structures or icons.
- Don't use local variables just to avoid long wires! Every local variable that reads the data makes a copy of it.

- Reduce the number of zigzags by aligning inputs and outputs. Use the cursor keys to take kinks out of your wires.
- Delete excess wires, such as loops.
- Try to keep parallel wires evenly spaced even around corners.

8.2.3 Labeling Standards

- Every frame in a Sequence, Case, or Loop structure should have a descriptive comment. These should be indented if there are embedded structures.
- Use free labels on long stretches of wire to label the data. Place the label right on top of the wire with a transparent background.
- Comments for wiring shall have a white background with no border. A bordered comment could be mistaken for an icon.
- When a VI is loaded on a different platform, the fonts adapt to that new platform. On the front panel, LabVIEW tries to move the labels so they don't overlap the controls. It does not do this on the diagram. It's best to place labels below the objects they describe, so when they grow and shrink on the bottom and right sides, they stay next to the object. If you place a label to the left of its object, right-justify it so that it grows to the left.
- Use the labels for Boolean cases to document the case states. This not only improves the readability of your code but it also helps with a program's logic. You will find that the simple act of documenting a logical case will stop you from getting tied up with the ORs, ANDs, and negative logic.
- Use the labels on While Loops and For Loops to document their purpose.
- Label shift registers with their data contents.
- For labels and constants use Size to Text and Carriage Returns for each line.
- Standardize on Fonts and Styles.
- Highlight comments with a yellow or pink background to emulate a highlighter pen. We don't use this but thought it was reasonably neat and worth suggesting.

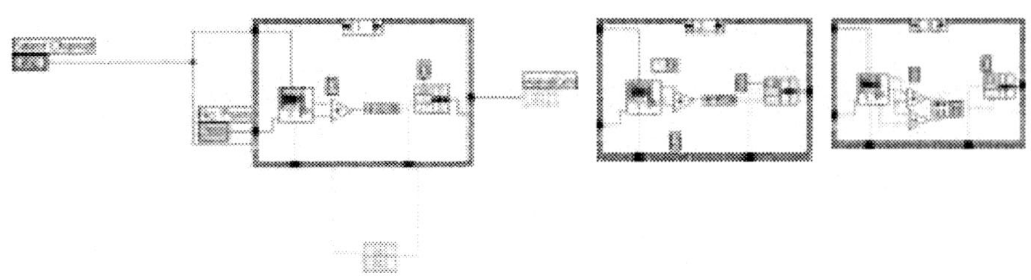

Figure 8.4
Can you guess what it is yet?

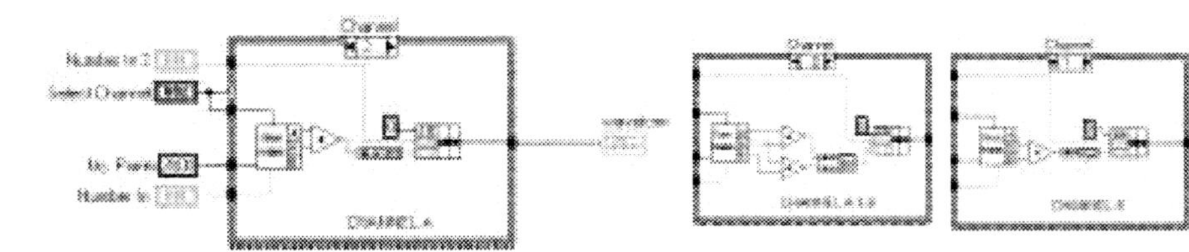

Figure 8.5
Aha, it's a scope.

8.2.4 Self-Documenting Example

The simple diagram in Figure 8.4 is not actually conveying any information, and there are also several uglies in there as well. A little bit of tidying up will work wonders for its readability. LabVIEW is an iconic language—spend a couple of minutes on an icon and it's time well spent. Enforce the left to right rule, add a few labels, and you'll witness a vast improvement, as Figure 8.5 demonstrates. If you do this from the start, it takes about the same amount of time to do it neatly as it does to make a mess of it.

By changing the Select Channel U32 to an enumerated type we can make the case statement more descriptive. We can also apply clear constraints on the Channels as a postcondition (you can also do it using the data range option for a number, but it isn't particularly visible and can lead to unpredictable program behavior).

Hey, where are all the labels in Figure 8.6? If you think about it, the typing of labels is a weakness. It's extra work that doesn't add anything to the program. It is purely there for the reader. If there is complex logic or algorithms use labels to explain them, but the use of enumerated types in your programs actually reduces the need for manual labeling.

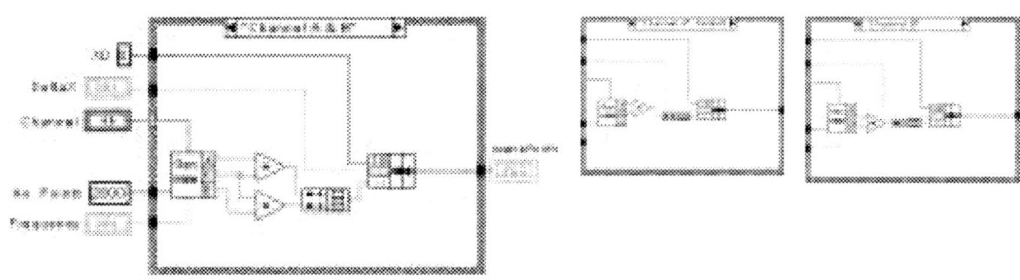

Figure 8.6
Get rid of the constants and we'll be in business.

8.3 Front Panel

8.3.1 General Front Panel Standards

All we're going to say about fonts is to use a standard font. Hey, we're all grown-ups, use whatever you like, we don't care! For some reason Times New Roman invokes a negative response from customers; they don't like the flourishes or something like that. It's not a bad idea to standardize somewhat on labels, but don't lose too much sleep over this.

The context help in LabVIEW is an excellent place to put all information that is relevant to, and needs to be kept with, the VI. It's a good idea to have a template for the VI's description, keeping a common look and feel to the documentation. See the example in Figure 8.7.

8.3.2 Public Front Panel Standards

- Use Strict Type Definitions for Enumerated Commands and Attributes.
- Fill in the Icon. It only takes a minute, and because LabVIEW is an iconic language it will greatly improve the readability of your program.
- A user description for each control or indicator will aid in documentation of the application and will also make LabVIEW help facilities available to the user.
- Keep a common look and feel, at least for the same customer. This will improve familiarity and reduce the learning curve when the customer is supplied with new software.

```
[Ver x.xx]
DESCRIPTION
This should briefly describe the operation of the VI.

COMMANDS
For an enumerated type command structure, list the commands and their
operations, inputs and outputs.

PRE-CONDITIONS

POST-CONDITIONS

ERROR HANDLING
Does the VI Pass through errors, ignore or process them. If an error is
passed into will the VI operate?

-----------------------------------------------------------------------
VERSION CONTROL
List all changes from released version.
For Updates increment the suffix number (x.01 will update to x.02). For New
versions (Branches) Increment prefix number (1.xx to 2.xx).
If the VI is in development prefix the number
with D (D1.xx).
```

Figure 8.7
VI description.

8.3.3 Private Front Panel Standards

This applies to all panels of subVIs that are not available or visible to the
user.

- Connector Controls on the left
- Connector Indicators on the right
- Local Indicators and Controls in the middle or on the bottom
- Align and evenly space them
- Have them in the same order as they are connected
- Frame the relevant areas Input, Output, and Local where applicable
- Use and label the recessed frame from the decorations menu
- Use the brush with the foreground set to transparent to tidy up the label

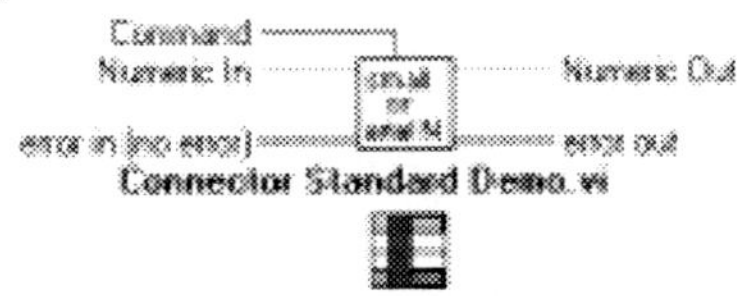

Figure 8.8
Connector standard.

8.3.4 Icon and Connector Standards

We've said it before but just to reinforce, if you can't afford the 60 seconds required to fill in the icon then your project is in real trouble!

- Fill in your icons.
- A readable and compact font is Arial Narrow at size 14. Just double-click on the letter "A" icon in the Icon Editor.
- Small Fonts at size 11 work well as an alternative.
- Place Inputs on the right, Outputs on the left, and commands up the middle.
- If possible, for full-size icons, keep to 4 in and 4 out as a connector pattern. This helps to keep your wiring tidy.
- Wire error clusters along the bottom input and output connector.
  The standard connector pattern that we employ is shown in Figure 8.8. The component command can be wired top or bottom, the error cluster is at the bottom inputs and outputs.

8.3.5 Organization of Files

It's reasonable to organize your files into a structure that emulates your software architecture. We also tend to declare public and private directories.

What do we mean by public and private? A public VI is a VI that hasn't got a specific owner, whereas a private VI has only one caller. This helps to encourage information hiding because the private subVIs are of no interest to any other developer and are clearly stated as such. Figures 8.9 and 8.10 demonstrate this relationship.

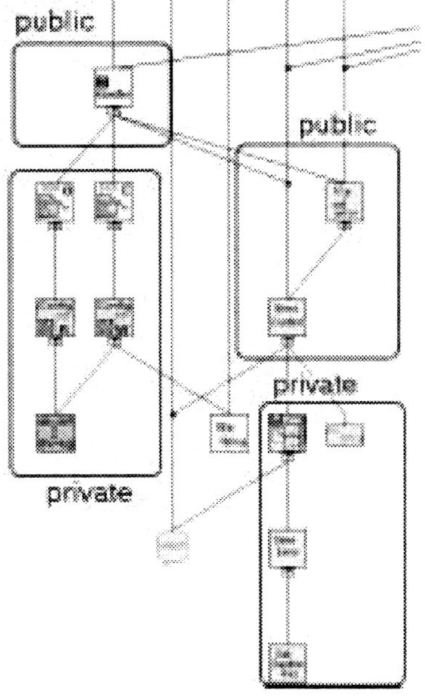

Figure 8.9
Hierarchy.

Figure 8.10
How it maps to directories.

For all the standards mentioned above, a reasonable way of enforcement is to provide a checklist. You could run through the checklist prior to releasing the component. Another excellent method of encouragement is the code review. Sitting down and having to justify your design and style decisions before your peers is one of the most powerful software quality methods available to you.

Remember though that standards should be agreed and changed by consensus. Life is too short to become a "Style Fascist."

The Journey

9

The idea of a journey is all very Zen, but we feel that it describes the fact that most engineers view self-improvement as a large motivational factor. The overall destination may be to build the perfect software and to be loved and admired by our peers. Along the way we have smaller destinations—finish a project, learn a new software tool, improve our productivity, and reduce stress levels. We also like the idea of having fellow travelers. Another important consideration we would like to emphasize is that we are still on our journey, and "perfect" is a very elusive destination, so any help from our fellow travelers would be greatly appreciated.

A project is like a journey as well; some projects can be akin to traveling uphill in mud! Successful projects have a clear destination, a carefully planned route, a map, a compass, and recognizable milestones along the way. Unsuccessful projects have no destination, an unclear destination, or a destination that moves faster than you do. Unsuccessful projects also have inadequate and inaccurate directions or, in the worst case, no directions at all.

This chapter will put meat on the bones of the previous chapters by applying what we've been discussing to an example project. We'll take our project from the very start and discuss along the way the common issues encountered. We'll demonstrate how to apply the techniques discussed in the preceding chapters. Along the way we'll highlight the areas where projects often go bad and discuss why. The demonstration software used in this chapter could be used as a pattern for your own project. We've certainly used it for ours.

Most projects start with an inquiry and an initial meeting to discuss your customers' requirements. Even if you are working internally within a company it doesn't exclude you from having customers such as your manager or the end user, for example.

9.1   Agreeing on the Destination (Requirements)

First, get down to basics by listening to what your customer wants and then telling the customer what to expect. Each party understands exactly what the other party is talking about.

To get the ball rolling then, listen to what your customer wants. Right from the start the project can take a wrong turn, your customer could call you and you could do everything over the phone. Following is a list of ways you could undermine your project's chances of success at an early stage:

- Not having requirements
- Having self-written requirements
- Having requirements written by the wrong person
- Not identifying the correct customer representative

The customer should have an idea about the destination, and before setting out you should be able to confirm with the customer at least the general direction. So, if you're lucky there will be a requirements specification. Here's an example.

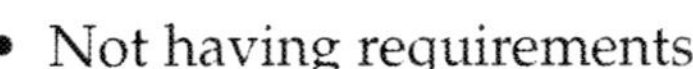

ACME WIDGET CO.

Description	Widgetometer Test System Requirements Specification
Document Number	RS_WATE10001
Issue	1.00

1.0 INTRODUCTION

The ACME Widget Company produces world-class Widgets and is a market leader in all types of Widgety devices. The Widgetometer is a device for measuring the Widgetiness of the Widgets. The requirement is to produce a test system for automatically testing Widgetometers to the customer's test specification.

2.0 RELATED DOCUMENTS

Example of Widgetometer test specifications.

3.0 THE WIDGETOMETER

The Widgetometer is the new way of measuring Widgets in the field; each Widgetometer can measure two Widgets for their Widgetiness. The Widgetometer communicates by its displays, two 0–10 V output ports, and RS232 communications.

See the following schematic:

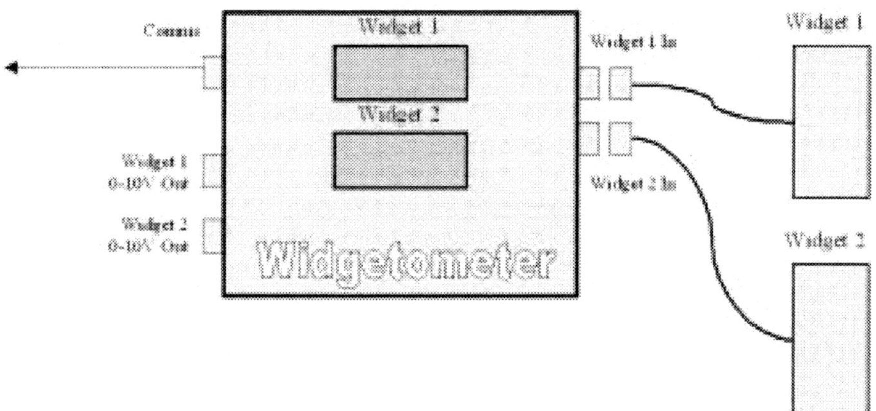

4.0 THE TEST SYSTEM

The system will test up to five Widgetometers against two gold-standard Widgets. The standard Widgets will need to be connected to both of the Widgetometer Under Tests ports. The requirement will be to take a reading from the Widgetometer's communication port and both of the Analog output ports. There is also a requirement to conduct a visual inspection of the display to confirm that the readings all agree. The standard Widgets will be calibrated, and the readings should agree with this calibration. The results will be recorded and stored in a retrievable format, and a printout of the test will be provided.

5.0 User Input

The user will plug the units into their test position and press a button to start the test. The units under test will identify themselves via the communications port. All the active test positions will then be tested to the customer's test specification for the part number.

6.0 System Output

On completion of a test, the UUT (unit under test) data will be archived. On completion of all the tests a printout will be produced.

This document is usually accompanied by an initial requirements meeting. If no requirements document is made available prior to this, one should be discussed and produced from this meeting.

AN EXAMPLE REQUIREMENTS MEETING

The meeting was held at the customer's premises and the following people were present:

Ken Customer—Project Engineer

Millicent Manager—Ken's Boss

Derek Developer—The Software Engineer

Brief introductions are made and now we'll go straight to the nitty-gritty:

Ken: The Widgetometer is the new way of measuring Widgets in the field; each Widgetometer can measure two Widgets for their Widgetiness. The Widgetometer communicates by its displays, two 0–10 V output ports, and RS232 communications.

Millicent: We've not finalized the protocol for communicating through the RS232 yet. This is due shortly.

Ken: As you can see the product is still being developed, unfortunately, the development cycle ends a very short time before the product goes into production. We'll try to keep nasty surprises to a minimum.

Millicent: LabVIEW is flexible enough to cope with any late changes, right?

Now Derek has got to awake from his slumber and respond to the question from Millicent. He has various options; here are a few:

1. **Bluff.**

    Yeah, LabVIEW can cope with anything you throw at it.

    The problem here is that a bluff nearly always gets overtaken by reality. The other problem is that you are giving the customer the go-ahead to make changes without any consideration. Go this way, and you'll end up having end of project blues.

2. **Be strict.**

    We will need a complete specification before we can begin. Changes later on are far too expensive to rectify and will weaken the design of the software.

    For languages like C or Pascal this is a reasonably valid path to follow, because if using these languages, the developer will disappear for three months and reappear with a finished product. This can be a recipe for customer disappointment if the requirements are not concisely and accurately defined and understood by all concerned parties. LabVIEW gives the advantage of Rapid Prototyping, use it!

    You also have to consider that the only constant in software projects is change. As your customers learn more about the system they will have more suggestions and comments. It's a fact of life and we all have to cope with it. A large amount of work will pass you by with this attitude because in business customers don't want to wait.

    Finally, a decent language combined with a robust design strategy will allow flexibility at a reasonable cost.

3. **Tell it like it is.**

    LabVIEW is the language for writing the application, and the design of the application can be flexible or not. This depends on the effort put into the design and the inherent complexity of the system. We should design the system with flexibility in mind, but obviously there will be a cost consideration for large system changes. As long as the basic requirements stay the same we can reevaluate the process at each milestone.

We will also need to ensure that changes in requirements are communicated in an efficient manner and that only one person in your organization communicates them to us.

Go for it Developer man!

Anyway, we've chosen option 3 and the meeting continues.

Ken: As described in the requirements specs, the system we need will test up to five Widgetometers against two gold-standard Widgets. The standard Widgets will need to be connected to both of the Widgetometer Under Tests ports. We will want to take a reading from the Widgetometer's communication port and both of the Analog output ports. Finally, we will need to conduct a visual inspection of the display to confirm that the readings all agree.

Derek: Will there be a limit applied to the readings taken?

Ken: The standard Widgets will be calibrated and the readings should agree with this calibration. Perhaps if I get an example of a customer test specification that would help.

Derek: Right, before we finish we need to establish a project representative. This should be a customer who is assigned the overall responsibility for the project. We will use this person for all design decisions and approvals.

The next step after the requirements have been communicated, is for you to tell what the customer is going to receive. This could be in the quote, contract, or as we like to call it, the target specification, and it should be regarded as your promise to the customer.

This is a communication-intensive process, so be prepared to talk to your customer. For example, it's a rare event when you get a requirement specification with no gaps in it. In this case there is little mention of the data to be collected by the system. Call them and clarify before you quote!

DEREK'S SOFTWARE CO.

Description	Widgetometer Test System Target Specification
Document Number	TS_WATE10001
Issue	1.00

1.0 INTRODUCTION

This is a response from Derek's Software Co. to ACME Widget Company document number RS_WATE10001 Issue 1.00.

That document expressed the need for a system to automatically test Widgetometers to a customer's test specification.

2.0 RELATED DOCUMENTS

Requirements Specification RS_WATE10001 Issue 1:00
Schematic of Widgetometer Test System DRG_WATE10001001 Issue 1:00

3.0 SOFTWARE, OPERATING SYSTEM, AND HARDWARE

The system will be written in LabVIEW version x.y and will work on Windows 2XXX operating systems. All source code will be provided as part of the contract.

The software will need a minimum of the following to run satisfactorily:

COMPUTER

> Bloatium III processor
>
> 64Mb RAM
>
> 10Gb hard drive
>
> 1,024x768 SVGA true color display
>
> CD-ROM
>
> 5 PCI slots

Associated plugs, cables, and wiring do not fall into the scope for this target specification.

MEASURING SYSTEM

> DC Voltage Accuracy of better than ±0.5% and a resolution better than 1 μV.

COMMUNICATIONS

Communication will be RS232, 7 or 8 data bits, 0 or 1 start and stop bits, odd, even, or no parity. Data transfer rate will be between 2,400 and 19,200 baud inclusive. Protocol to be established.

4.0 USER INTERFACES

The main display will contain all of the information from the system. Control buttons will be along the bottom of the screen. [Exit] stops and unloads the program. [Start Test] will commence testing. A dialog will be displayed to allow the user to enter any data that cannot be taken automatically from the units under test. The test sequence will be quick enough to remove the necessity of an abort test function. Test stage details will be displayed in the central display area.

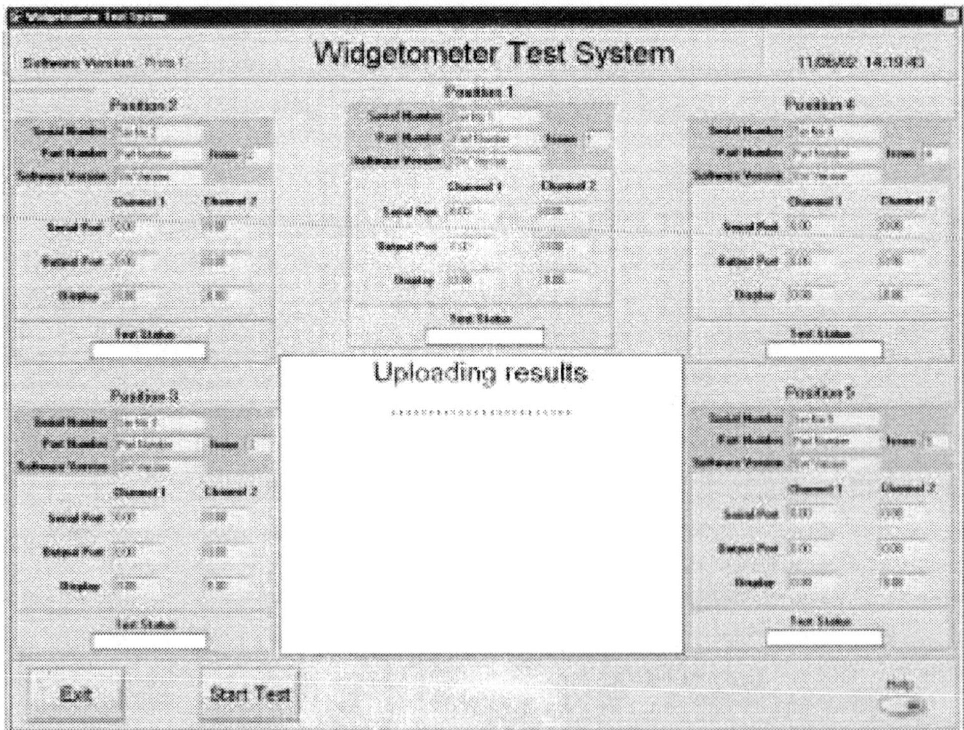

5.0 TEST SCENARIO

There will be a button that will allow the user to log in and out of the system. [Exit] and [Start Test] will be disabled when no one is logged in. When testing, the [Login] button will be disabled.

The units will be loaded into the test system and [Start] pressed on the main screen. The test type dialog will be displayed and a test type selected or new test type entered. The test type will define the limits applied to each test.

The unit details will be taken from the serial ports of all the units available, and any test positions that do not respond will be ignored. The units will then be connected in order to the two standard Widgets and their readings taken from the serial ports as well as the analog ports. A dialog will also be displayed that will request the readings from the Widgetometer displays.

The data for each test will be stored on the completion of that unit's test. When all the tests are complete a test report will be printed.

6.0 TEST REPORT FORMAT

User Name

Date of Test dd/mm/yy

Unit 1—Serial Number xxxxx	**Time of Test** hh:mm:ss
PartNumber ppppp	**Issue** x.xx
Software Version x.xx	**Test Type** 10%Test

Reading Type		Actual		Read		Error		Status	
An Out 1	|	1.000	|	0.95	|	±10%	|	PASS	|
An Out 2	|	2.000	|	1.95	|	±10%	|	PASS	|
Serial 1	|	1.000	|	0.89	|	±10%	|	FAIL	|
Serial 2	|	2.000	|	1.78	|	±10%	|	FAIL	|
Display 1	|	1.000	|	0.95	|	±10%	|	PASS	|
Display 2	|	2.000	|	1.95	|	±10%	|	PASS	|

Unit 2—No Response

Unit 3—No Response

Unit 4—No Response

Unit 5—Serial Number   xxxxx   **Time of Test**   hh:mm:ss

PartNumber   pppppp   **Issue**   x.xx

Software Version   x.xx   **Test Type**   10%Test

Reading Type		Actual		Read		Error		Status	
An Out 1	|	1.000	|	0.95	|	±10%	|	PASS	|
An Out 2	|	2.000	|	1.95	|	±10%	|	PASS	|
Serial 1	|	1.000	|	0.89	|	±10%	|	FAIL	|
Serial 2	|	2.000	|	1.78	|	±10%	|	FAIL	|
Display 1	|	1.000	|	0.95	|	±10%	|	PASS	|
Display 2	|	2.000	|	1.95	|	±10%	|	PASS	|

7.0 STORED DATA FORMAT

Data Stored . . .

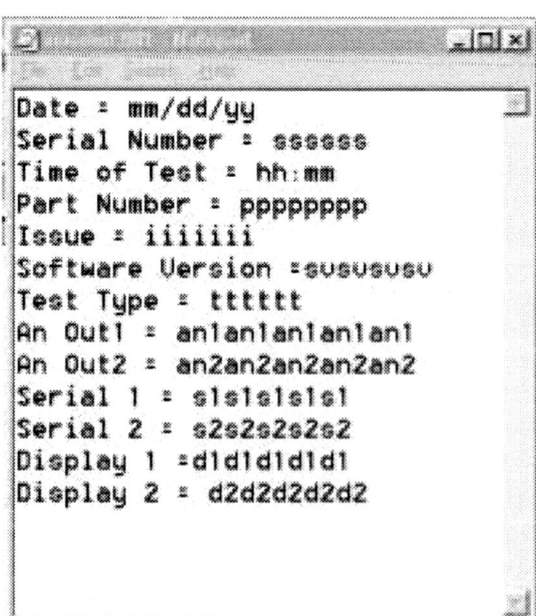

The file will be Comma Separated Text File, and the directory structure and name will correspond to unit serial number and date.

Widgetometer Results
|_ __ Part Number
    |_ __ Year

8.0 PROJECT MILESTONES AND RESULTS

week 1—Order Accepted

week 3—Initial Design Review

Design Review Document, Detailed Test Plan, Minutes

week 5—Final Design Review

Design Review Document, Detailed Test Plan, Minutes

week 15—Alpha Version

Customer review of software at developer premises

week 17—Beta Version

Beta release of software installed at customer premises

week 20—Full Customer Acceptance

Full release of software

Accompanying the target specification should be the test plan. This will be the instrument for confirming that the destination has been reached.

DEREK'S SOFTWARE CO.

Description   Widgetometer Test System Test Plan

Document Number   TP_WATE10001
Issue   1.00

1.0 RELATED DOCUMENTS

Requirements Specification RS_WATE10001 Issue 1:00

Schematic of Widgetometer Test System DRG_WATE10001001 Issue 1:00 (Figure 9.1)

Target Specification TS_WATE10001 Issue 1:00

2.0 ACCEPTANCE TESTING

These tests are to confirm the system is satisfactory to the customer's requirements document (RS_WATE10001 Issue 1:00).

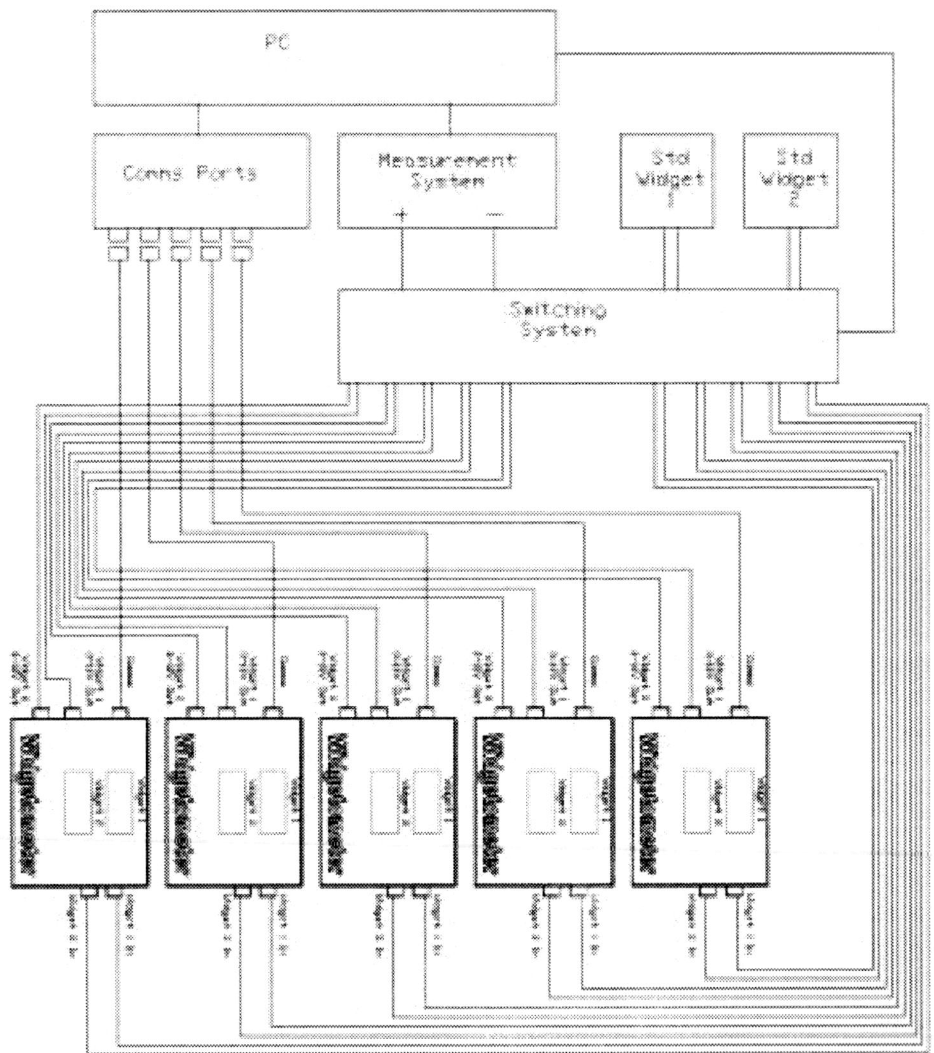

Figure 9.1
Test system schematic.

2.1 TEST EACH POSITION FOR A UNIT LOADED AND A UNIT NOT LOADED

Take a unit with known attributes and load into position 1 and run the test. Move to position 2 and repeat, move to position 3 and repeat, move to position 4 and repeat, and move to position 5 and repeat. Confirm that the loaded tests agree with known values and the unloaded tests report an error.

Action	Result	Pass/Fail
Load Position 1		
Load Position 2		
Load Position 3		
Load Position 4		
Load Position 5		
Unload Position 1		
Unload Position 2		
Unload Position 3		
Unload Position 4		
Unload Position 5		

2.2 TEST MEASUREMENT QUALITY AND REPEATABILITY

Take five units with known attributes and fully load the test system and run a test. Confirm that the tests agreed with known values.

Repeat the tests five times and confirm that readings do not drift beyond acceptable limits.

Unit	Serial Number	Chan 1 Output Range	Chan 2 Output Range	Chan 1 Comms Range	Chan 2 Comms Range
1					
2					
3					
4					
5					

2.3 TEST EACH CRITERIA FOR A PASS AND FAILURE CONDITION

For positions 1 and 5 set the limits to fail on all the parameters and then all the parameters individually.

Action	Result	Pass/Fail

Load unit xxxx into position 1 and set limits to fail on . .
    Chan 1 Output
    Chan 2 Output
    Chan 1 Comms
    Chan 2 Comms
    Chan 1 Visual
    Chan 2 Visual		

Load unit xxxx into position 5 and set limits to fail on . .
    Chan 1 Output
    Chan 2 Output
    Chan 1 Comms
    Chan 2 Comms
    Chan 1 Visual
    Chan 2 Visual		

Load unit xxxx into position 1 and set limits to fail on . . Chan 1 Output		
Load unit xxxx into position 1 and set limits to fail on . . Chan 2 Output		
Load unit xxxx into position 1 and set limits to fail on . . Chan 1 Comms		
all tests in here	↓	↓
Load unit xxxx into position 5 and set limits to fail on . . Chan 1 Visual		

Action	Result	Pass/Fail
Load unit xxxx into position 5 and set limits to fail on . . Chan 2 Visual		

2.4 CONFIRM FILE FORMAT OKAY

Action	Result	Pass/Fail
Open generated files and confirm that they agree to the format described in the target specification		

2.5 CONFIRM FILE DIRECTORY STRUCTURE OKAY, ESPECIALLY AT THE BOUNDS OF MIDNIGHT, MONTH, AND YEAR CHANGEOVERS.

Confirm that the file directory structure agrees with the system time and date.

By adjusting the system date and time confirm that new directories are generated correctly for the midnight changeover, month changeover, and year changeover.

Action	Result	Pass/Fail
Midnight Changeover Before, Set time to 11:50 pm and run test		
Midnight Changeover After, Set time to 12:01 am and run test		
Month Changeover Before, Set date to Sept 30 2002, Set time to 11:50 pm and run test		
Month Changeover After, Set Date to Oct 01 2002, Set time to 12:01 am and run test		

Action	Result	Pass/Fail
Year Changeover Before, Set Date to Dec 31 2002, Set time to 11:50 pm and run test		
Year Changeover After, Set Date to Jan 01 2003, Set time to 12:01 am and run test		

2.6 Confirm Printouts Okay

Check that the details on the printout agree with the data taken and that the format corresponds with the format described in the target specification.

Action	Result	Pass/Fail
Printout agrees with data taken. Attach printout and hard copy of file.		
Format of printout agrees with target specification TS_WATE10001 Issue 1:00		

Finally, we suggest the use of a prototype to better elicit requirements. However, this is usually only realistic at the requirements stage if the project has a separate cost feasibility stage. Alternatively, this can also be done as part of the design process to fine-tune the requirements. For your own sanity be prepared. Even if your software eradicates poverty from the world there will be someone who complains about the font!

Without a target specification your customer will not have confidence in what he or she is getting, and without a test plan the project will have no closure. You know what you are producing and your customer knows what he or she wants. A project will only be successful if everyone agrees.

So to summarize, the target specification should describe the key functionality of the system, and the test plan should list the tests that will establish that this functionality has been met.

In our journey these documents are the backpack, walking boots, food, compass, and map.

9.2 Planning Your Route (Design)

9.2.1 Code and Fix

Who needs design anyway. LabVIEW makes it easy for us to throw a few indicators on the front panel and use the lasso to squish our code down to acceptable proportions.

NO NO NO NO NO NO NO NO NO NO NO NO!

This is like embarking on a journey armed with unclear directions, no map or compass, and a vague idea of your destination.

We don't suggest you go down this road, since death and pestilence will be your traveling companions.

9.2.2 Abstracting Components from Requirements

The first thing to do is to trawl the available information for candidate components. OO texts tell you to do a noun-verb parse. Nouns can be used to define a component or a component attribute, and verbs are generally actions that are applied to components. English is a very imprecise language full of innuendo and subtle meaning. So, the point we are aiming for is to get an understanding of the test system and for this understanding to be reflected in the model.

Noun Parse of Requirements Specification

Noun	Noun
Widget	Analog Output port
Widgetometer	Visual Inspection
Test system	Calibration

Noun Parse of Requirements Specification (continued)

Noun	Noun
Test specification	Test position
Display (Widgetometer)	Active test position
Output Port (Widgetometer)	Customer test specification
RS232 Communications	Part number
Standard Widget	Data
Test ports	Printout
Reading	UUT
Communication port	

Noun Parse of Target Specification

Noun	Noun
Main Display	Part number
Central Display Area	Time of Test
Control Buttons	Issue
Dialog	Software Version
Units Under Test	An Out
Test Type	Serial Out
Unit Details	Display Out
Serial Ports	File
Test Report	Directory Structure
User Name	
Date of Test	
Serial Number	

There are a few hidden components that most systems will need. For example, for this ATE there is no mention of the display and how it should look. A phone call confirms that the customer would like to see the test status and unit details on a screen.

We can also have a look at the hardware design to swipe a few ready-made components. For example, there is a digital multimeter (DMM) for taking the measurements, a Switching System for making the connections, and a PSU (power supply unit) for providing power. Now you can try to disregard repeats and obvious components that are outside the system.

So, let's go through our list and collate what we have identified into a smaller set of nouns, deleting those that we don't need:

Nouns	
Widget	
~~Widgetometer~~	*Discard—is not meaningful for ATE*
Test system	
~~Test specification~~	*Discard—not part of system*
~~Display (Widgetometer)~~	*Put in Hardware*
Output Port(Widgetometer)	
RS232 Communications	
Standard Widget	
Test ports	
Reading	
Communication port	
Analog Output port	
Visual Inspection	
Calibration	
Test position	
~~Active test position~~	*Discard—replace with Test Position*
~~Customer test specification~~	*Discard—not part of system*

(continued)

Noun	Noun
Part number	
Data	
Printout	
UUT	
~~Main Display~~	*Replace with Display*
~~Central Display Area~~	*Replace with Display*
~~Control Buttons~~	*Replace with Display*
~~Dialog~~	*Replace with Display*
Units Under Test	
Test Type	
Unit Details	
Serial Ports	
Test Report	
User Name	
Date of Test	
Serial Number	
Part number	
Time of Test	
~~Issue~~	*Discard—relates to documentation*
Software Version	
~~An Out~~	*Discard—same as Analog Output Port*
~~Serial Out~~	*Discard—same as RS232*
~~Display Out~~	*Discard—same as Widgetometer display*
File	
Directory Structure	

From our table we can see that some nouns have been discarded, but others have been replaced with (we think) a more meaningful description—Display.

Customer test specification and test specification are obviously referring to the same thing, so drop customer test specification because the system will not care who owns the specs.

There's also an argument that test position and active test position are similar enough to be ignored. Active could be an attribute of test position perhaps.

All right, our next job is to group the nouns under a heading.

Data	Hardware	Test System	Report	Display
Test Report	Widget	Visual Inspection	Printout	Display
User Name	Output Port	Calibration		Control Buttons
	(Widgetometer)			
Date of Test	RS232 Communications	Test position		Dialog
Serial Number	Standard Widget	Part number		
Part number	Test ports	UUT		
Time of Test	Reading	Units Under Test		
File	Communication port	Test Type		
Directory Structure	Analog Output port	Unit Details		
Software Version	Test position	Serial Ports		
	Widgetometer display			

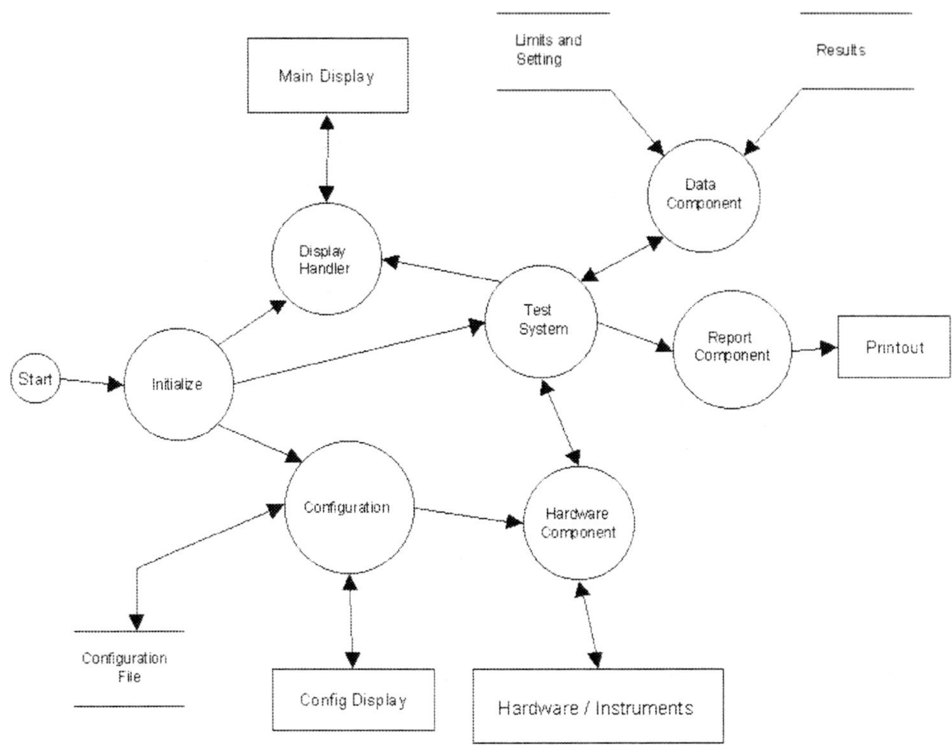

Figure 9.2
Widgetometer top-level component interaction diagram.

We have grouped each of our nouns under a heading that collectively describes it. This should be our starting point for the highest level components. More than one design can solve a problem, and you could have dissected the nouns differently and still come up with a decent design. For our example though, we have decided that our table headings now represent the main components in our system.

Putting it all together we will have a display that is populated by readings taken from the test system component. This test system will consist of a hardware component that will model all the states of the hardware. The hardware component will be responsible for talking to the serial communications, the DMM, and the visual inspection system. The results will be stored by a data component and printed by the report component. The component interaction diagram will look like Figure 9.2.

Now at the top level we can define the actions for each component.

Display Handler

This component describes the abstract things that the display will be required to do.

> Reset Display
> Show Test Position
> Update Test Position
> Update Status Area

Test System

This will be a state machine that will run through each test for each unit, checking pass-fail criteria and distributing the results data. So as well as the individual tests there will be states that model initialization, storing data, checking status, error states, and printing reports.

Hardware Component

Any individual action for the hardware is modeled by this component. This is at the highest level of abstraction.

> Initialize
> Measure UUT Channel 1 An Out
> Measure UUT Channel 2 An Out
> Read UUT Channel 1 Serial
> Read UUT Channel 2 Serial
> Read UUT Data Serial (this is Serial Number, Part Number, etc.)
> Get UUT Channel 1 Display
> Get UUT Channel 2 Display

Report Component

Any output to any printer is handled by this component. At this stage there is only one report required.

> Initialize
> Print Report

Data Component

All test data is held here. This component has the responsibility for loading test limits and settings as well as storing and archiving the test results.

Command	Attribute
Initialize	Measure Error %
Load Test Settings and Limits	Read Error %
Store Test Results	Display Error %
Get Limit	Measured Value 1 (An out)
Set Result	Measured Value 2 (An out)
	Read Value 1 (Serial)
	Read Value 2 (Serial)
	Read Value 1 (Display)
	Read Value 2 (Display)

Configuration Component

This component handles anything to do with the configuration data. The attributes will be set during development.

Load All
Save All
Load
Save
Get
Set

Initialization

This component sets up the configuration data and anything else that needs initializing.

At this stage the top-level components are doing very broad, abstract operations. The language of the commands is very specific to the test system. The components that these top-level components employ will become more and more specialized the farther down the program hierarchy they are.

Okay, this is a bit contrived, and since design is a gray area there will be other ways of doing it. Practice has shown us that using these techniques does throw up similar designs again and again. Patterns are a way of taking advantage of these prefabricated designs.

9.2.3 Using Patterns to Help the Design Process

System design should ideally be a collaborative effort between the developer and the customer. Lucky us, we're armed with our design patterns, which will allow us to ask very specific questions. The responses to these questions can be directly applied to the components in the pattern.

UI

There are two design patterns that we can employ here. If it looks as though the system is going to have a lot of individual displays, then we suggest starting off with the Menu Tree Pattern. If any of the displays are complex and require information from subVIs to be passed and displayed, we suggest using the UI Controller>>Message Queue Pattern for each.

This system doesn't require a large number of displays and therefore doesn't merit a menu-submenu system, but the complexity of the data to be displayed on the main UI does warrant the use of the UI Controller>>Message Queue Pattern.

First, let's reach a consensus about what the general UI operation is going to be.

A meeting is called with the main protagonists attending:

Derek: I've got quite a few ideas, concentrating purely on the User Interface. So let's take it in sequence from the beginning. The first thing you would see is a screen with buttons, away from the main display area, and some sort of representation of the test system.

(Derek now sketches his concept).

Derek: The five test positions will be displayed on the screen grayed out, one of the control buttons will have Start Test written on it, and a central area of the display will have instructions for this stage of the program. When Start Test is pressed the Test Setup dialog box will be displayed, and from this the required test position activated and the data for that test position entered. Upon satisfactory data entry, the dialog is exited and the test started. Are we all happy with the general flow so far?

Ken: I would like to be able to cancel the data entry and return to the previous state, but I like it apart from that.

Derek: I'll ensure that all dialogs can be canceled as a general requirement. When the tests have begun the currently active test details will be shown in the middle, with the active locations enabled.

Millicent: Can we have some sort of indication of test status? You know, something like Pass, Fail, Not Tested, Testing.

Derek: No problem, we could use labeling or background colors. I'll put that in as a general requirement and we'll see how it looks on the prototype. When an individual test is complete do you want any specific posttest action, like print a report or detailed test status display, or do you want to carry straight on with the next test and leave all the posttest processing until then?

Ken: The test sequence will not be too long, so I think we should allow the test to finish and print its report then.

Millicent: OK, but I would like to see the data for the test stored away as soon as possible.

Derek: All right, I think that gives me enough for now. I'll just note it in the minutes and read it back to make sure we're completely happy.

There has been a lot of information floating about, so Derek now writes the following on his sketch.

UI States

Initialize>>Set Up Display>>Run>>Finish

These are fairly repeatable from system to system. The other common state for the system could be a general error state where the program goes if it all goes horribly wrong.

Action	Description
No Action	Just ignore
Disable Position	Disables the selected position
Enable Position	Enables the selected position
Position Visible	Makes the selected position visible
Position Invisible	Makes the selected position invisible
Update Position Data	Updates the data for the selected position
Update Instruction Display	Sends text to the Instruction display
Set to Tests Not Started	Sets the display to the not started state

The beauty of separating the actions like this is that you can design the basics of the system and implement them for the prototype without getting bogged down in detail. You know that the User Interface will react to the commands even though the actual contents of the Unit Data areas have not been established yet.

UI Attributes

Position Number (1 . . . 5)

This is a useful technique for simplifying the interface. Derek could have described the system fully by having lots of UI actions. For example, he could have had Disable Position 1, Disable Position 2, . . . Disable Position X. There is a potential problem in that the customer could come back and say that for cost purposes he or she wants to test 10 units. While not a showstopper, it could become cumbersome to have to write in all the commands for each unit.

Program Executive States

Test Not Started>>Enter Data>>Testing>>Tests Complete

There will also be a Test Executive defined as part of the Testing State. We'll come back to this when we have the Test Specification of the Widgetometer.

We now have an idea of the User Interface, so let's tackle the hardware.

Back to the meeting.

(Derek reads through his notes.)

Derek: Now let's concentrate on the specifics of the system. The first thing I would like to do is define what the hardware is doing in terms of the tests on the unit. What actions do we want to perform on the unit and what responses do we want to get back? Presumably we'll need to talk to the unit via its RS232 port. What actually do we want to tell the Widgetometer to do?

Ken: Well, I suppose we would want to retrieve its serial number, part number, and software version. Also, we would like to retrieve the Widgetiness of each Widget attached to the unit and the status of the unit.

Derek: We'll not worry about how to do it just yet, but just assume that these actions will need to be done. I'll take a quick note. What other tests will we be doing to the unit?

Millicent: We'll need to get the readings from the Analog Output port.

Derek: Just to prove I was paying attention, I'd suggest that we put the Visual Inspection Test in here. There's an argument that a manual visual inspection could eventually be replaced with an automatic one.

Hardware Actions

UnderRange

Read Unit Status

Read Unit Serial Number

Read Unit Part Number

Read Unit Software Version

Read Chan1 Reading

Read Chan2 Reading

Get Chan1 An Out

Get Chan2 An Out

Get Chan1 Display Reading

Get Chan2 Display Reading

OverRange

The Hardware component actions are now defined and there is a list of jobs that need doing. Using the Control >> Drive>> Read Pattern we can break down the components further (although this is not necessary for the purpose of the meeting).

Hardware Attributes

Unit Number (1 . . . 5)

Some customers haven't thought that deeply about what they want. It is also often the case that they don't fully understand what they want until they get it. This happens frequently, so we recommend the use of a prototype to demonstrate the User Interface interactions.

9.2.4 Building the Prototype

In LabVIEW, building a prototype is not necessarily expensive, and if you use patterns you can use the prototype as the basis for your main program. The idea of a prototype is to emulate the look and feel of how you envision the program working. Therefore, the User Interface should be functional to the extent that buttons can be pressed and data displayed. There shouldn't be any connections to the hardware at this stage. You shouldn't spend much more than one to two days putting the prototype together.

Another small bit of advice given to us by an ISO9000 auditor is "always feed them something," by this we mean deliberately leave something out or leave an obvious gap. In the auditor's case it was a ploy used by those being audited to distract. In our case we can use it to generate customer involvement, or to get customer "buy in" and start the conversion. We have found that people begin to accept the design more readily if they feel that they have contributed to it in some way. We should try to animate each step in the program—remember it is a demonstrator for the customer.

Step 1—Transfer of User Interface Pattern

Download and extract the examples from the Website. Copy the "UI Controller-Message Queue Pattern" directory in its entirety to a new directory called *"C:\Widgetometer\Software\."* In the directory, *\Widgetometer Test System\Software\1 Displays\1 Main Display*, change the name of the User Interface.vi to Widgetometer User Interface.vi. It is also a good time to change the Window Title for this VI.

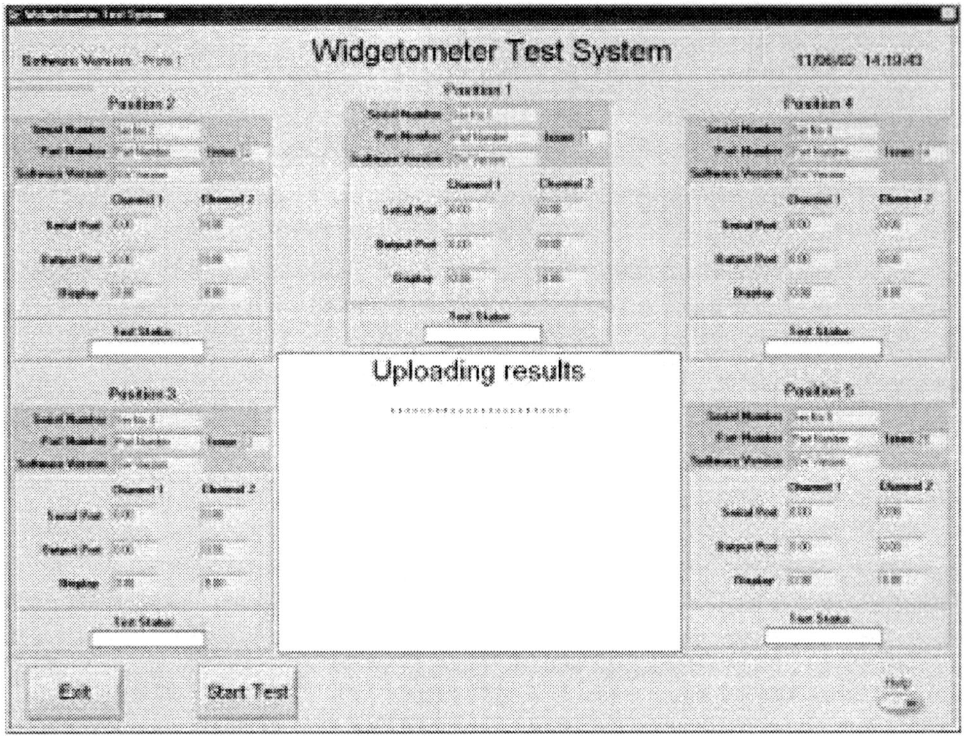

Figure 9.3
Prototype screen.

Step 2—Design your User Interface

> Dump all the controls and indicators onto the Front Panel and tidy up, as shown in Figure 9.3.

From the requirements document it was established that there would be five test positions. This was the limit of what could be displayed completely on the screen, without having to resort to summary screens. A Strict Type Def. cluster was created that held all the data for each unit under test. Copies of this cluster were distributed around the screen and titled Position 1–Position 5. Using a Strict Type Def. cluster will allow you to make changes to all of the clusters at the same time. The central area was allocated for instructions and test details and is just a plain string. Finally, the [Exit] button was left as is and [Button2] was renamed [Start Test].

Before we go on to actually emulating the running of the test, let's look at what basic actions we require from the User Interface.

User Interface Basic Actions

- Each position cluster will need to be updated, enabled, disabled, and made visible or invisible.
- For every [Start Test] button there should be a [Stop Test] button.
- The [Exit] button should be disabled during a test.
- [Button3] should be made invisible, since there is no apparent use for it yet.

Following are the User Interface Actions that came with the pattern:

Action	Description
Underrange	Throws or passes error
Clear Queue	Clears the queue of display actions
Get Display	Gets the next display item off the queue
Enable Button 1	as action
Disable Button 1	as action
Enable Button 2	as action
Disable Button 2	as action
Button 2 Visible	as action
Button 2 Invisible	as action
Enable Button 3	as action
Disable Button 3	as action
Overrange	as Underrange

Button 1 is the [Exit] button for our User Interface and the action should be updated to reflect this. Button 2 is the Start/Stop Test Button and, as well as renaming the button, we need further action to change the button text. Button 3 needs the visible-invisible actions.

We need to list new actions for the position clusters—these will be update, enable, disable, visible, and invisible. We should leave the UnderRange, OverRange, Get Display, and Clear Queue actions alone since they are always used.

The new User Interface Actions will need to be put into the Display Command.ctl and the Display Setting.ctl. Just open, edit, and save each control.

Display Command.ctl	**Display Setting.ctl**
Underrange Error	No Display
Clear Queue	Enable Exit Button
Get Display	Disable Exit Button
Enable Exit Button	Enable Start Stop Button
Disable Exit Button	Disable Start Stop Button
Enable Start Stop Button	Update Start Stop Button Label
Disable Start Stop Button	Enable Button 3
Update Start Stop Button Label	Disable Button 3
Enable Button 3	Button 3 Visible
Disable Button 3	Button 3 Invisible
Button 3 Visible	Enable Position
Button 3 Invisible	Disable Position
Enable Position	Position Visible
Disable Position	Position Invisible
Position Visible	Update Position
Position Invisible	Update Inst Screen
Update Position	
Update Inst Screen	
Overrange	

Create a Strict Type Def. enumerated control that models the test position, and its values should be:

Underrange, Unit 1, Unit 2, Unit 3, Unit 4, Unit 5, Overrange

Save it as Test Position.ctl and use it everywhere.

The UI queue component will have the Test Position.ctl as both input and output, and it will also pass it as part of the message queue. The adjustments to the UI queue component are relatively straightforward.

- Open the UI Queue.vi.
- Select the Test Position control using the "Select a Control . . ." button on the Control Palette.
- Drop it into the Input box and copy it to the Output box.
- Change the Test Position.ctl in the output box to an indicator.
- Copy and Paste the Test Position control into one of the Local Store Clusters. This is a Strict Type Def. and will need to be opened and the Test Position control pasted into it.
- Wire the input and output connector terminals.
- Finally, wire the input and output controls onto the block diagram.

The result should be something like Figure 9.4.

The cluster constants seen in the diagram in Figure 9.5 are instances of the Strict Type Def Queue Items.ctl, and any changes to it are reflected in changes to the diagram.

Figure 9.6 shows the UI controller component that acts as a wrapper for the UI queue component. Once again we'll need to add the Test Position.ctl as both input and output and wire them to the connector. If you open up the diagram, you'll notice that the source has changed to reflect the new values for the instances of Display Setting.ctl and Display Command.ctl. Spooky!

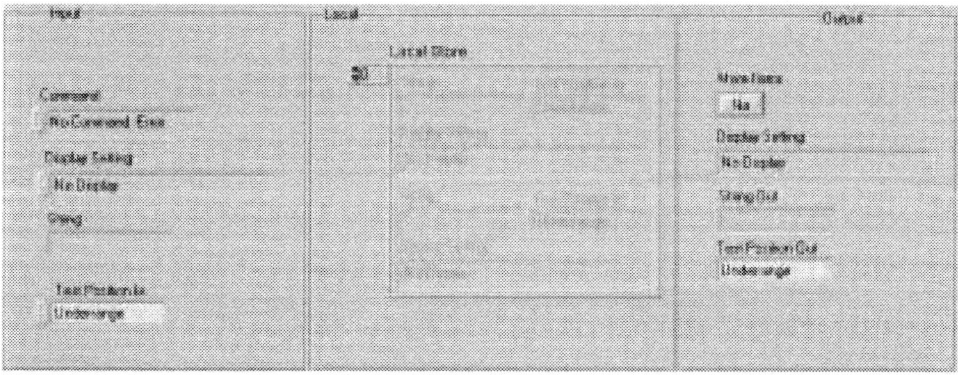

Figure 9.4
UI queue component panel and connector.

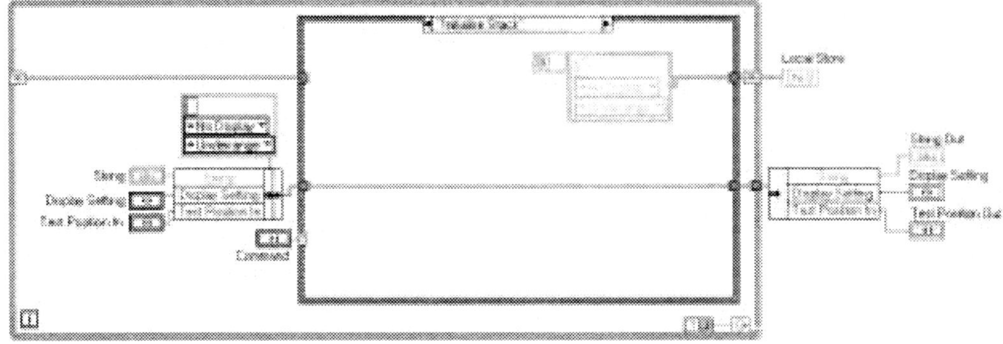

Figure 9.5
UI queue component diagram.

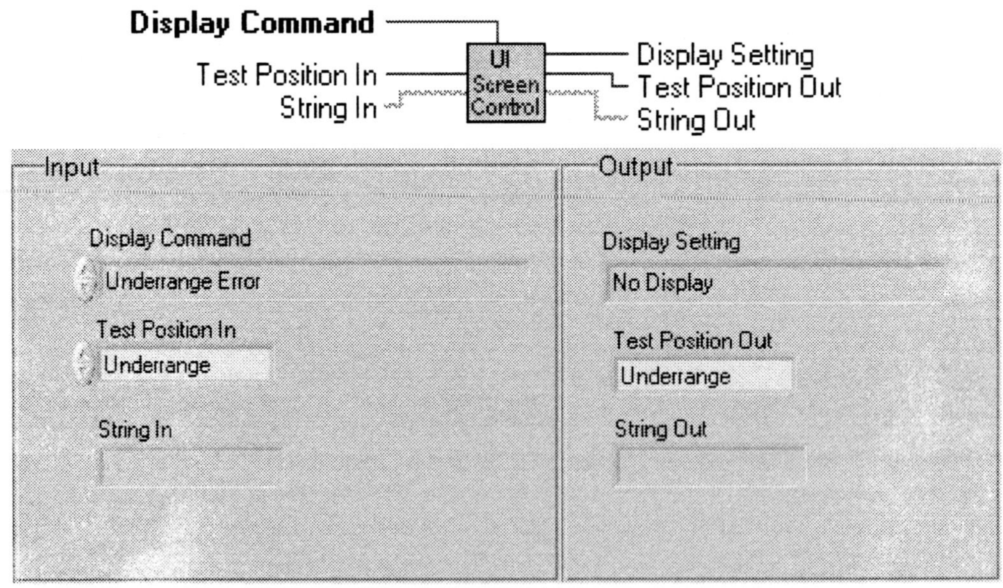

Figure 9.6
UI controller component panel and connector.

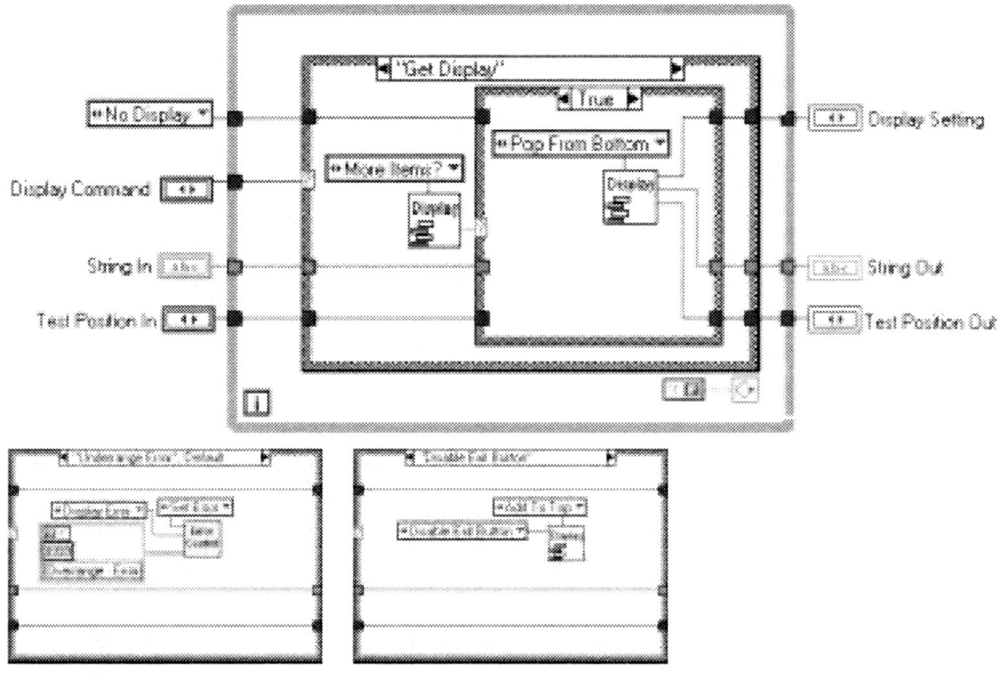

Figure 9.7
UI controller component diagrams.

The Test Position.ctl will need wiring into and out of the case structure and connecting to the UI queue component where required. In this case it is only required as an output from the Get Display>>More Items?=True case. See Figure 9.7 for the wiring diagrams.

Now we need to add the new cases to correspond to the elements in Display Command.ctl. This is a simple matter of selecting a suitable case and right-clicking and selecting *Duplicate Case* from the pop-up menu as shown in Figure 9.8.

The new case will automatically take on the name of the next element in Display Command.ctl, which in this case is "Enable Position." This case needs to be aware of the test position, so it will need Test Position wiring into the UI queue component. See Figure 9.9 for wiring of additional cases.

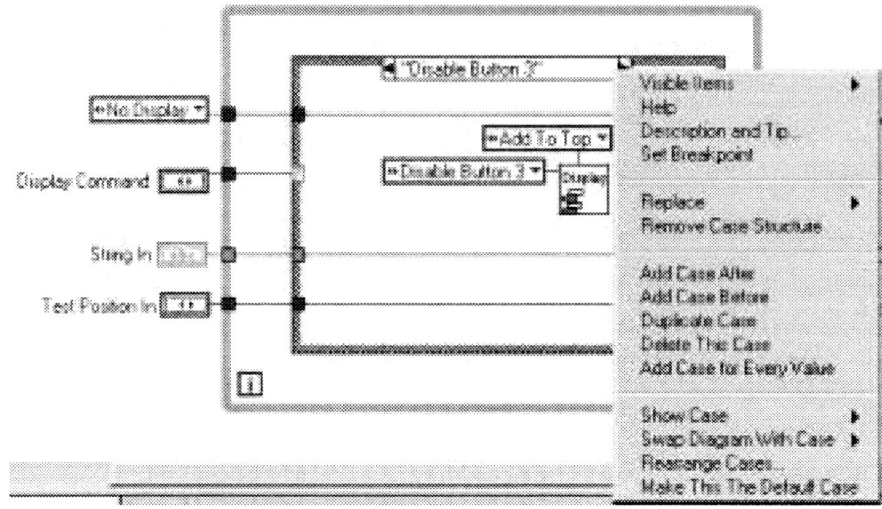

Figure 9.8
Duplicating a case.

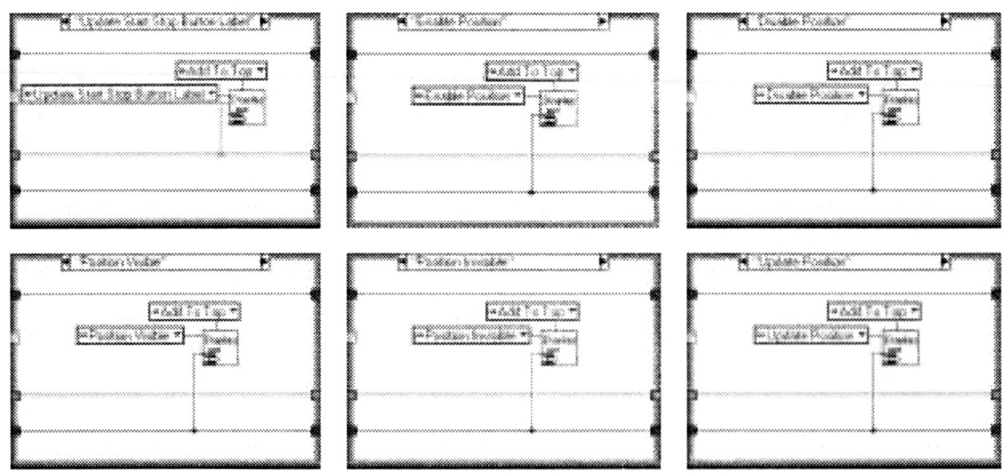

Figure 9.9
Additional cases for UI controller.

Finally, we need to make each action work! Look at the *Control Display While Loop* on the Widgetometer Test System block diagram in Figure 9.10. Right-click on the case structure and select *Add Case for Every Value.* Now you have a set of pigeonholes awaiting their actions. All of these actions are simple attribute settings, except for Update Position, and we'll tackle that later.

Here's the code for the Control Display While Loop as shown in Figure 9.10.

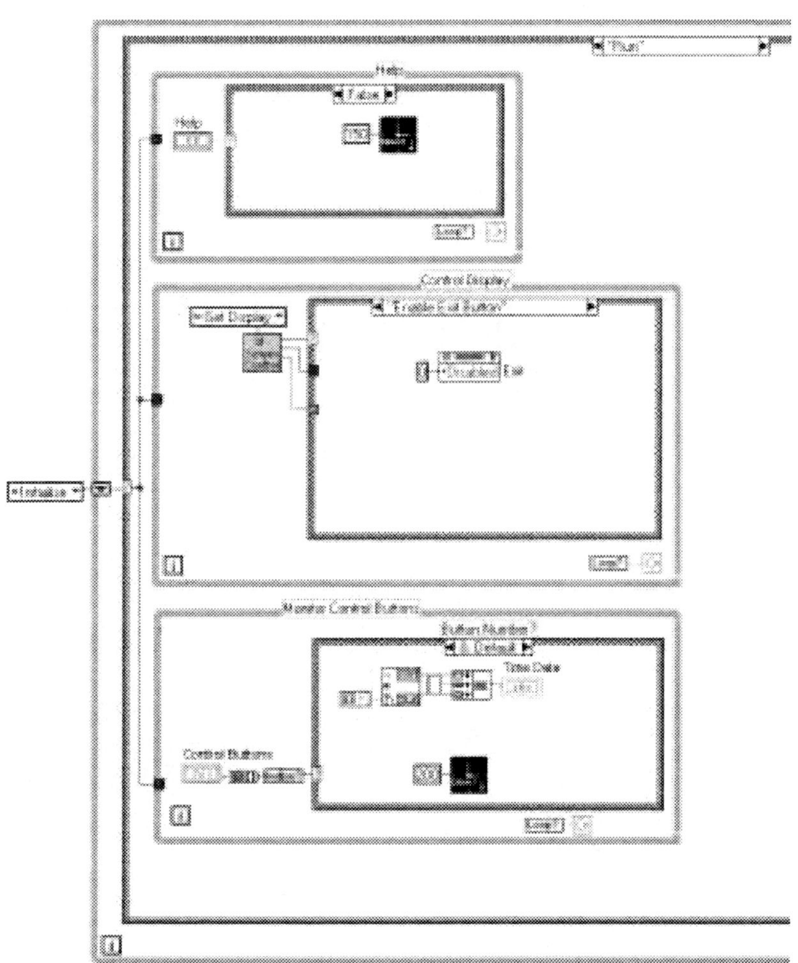

Figure 9.10
Control Display While Loop.

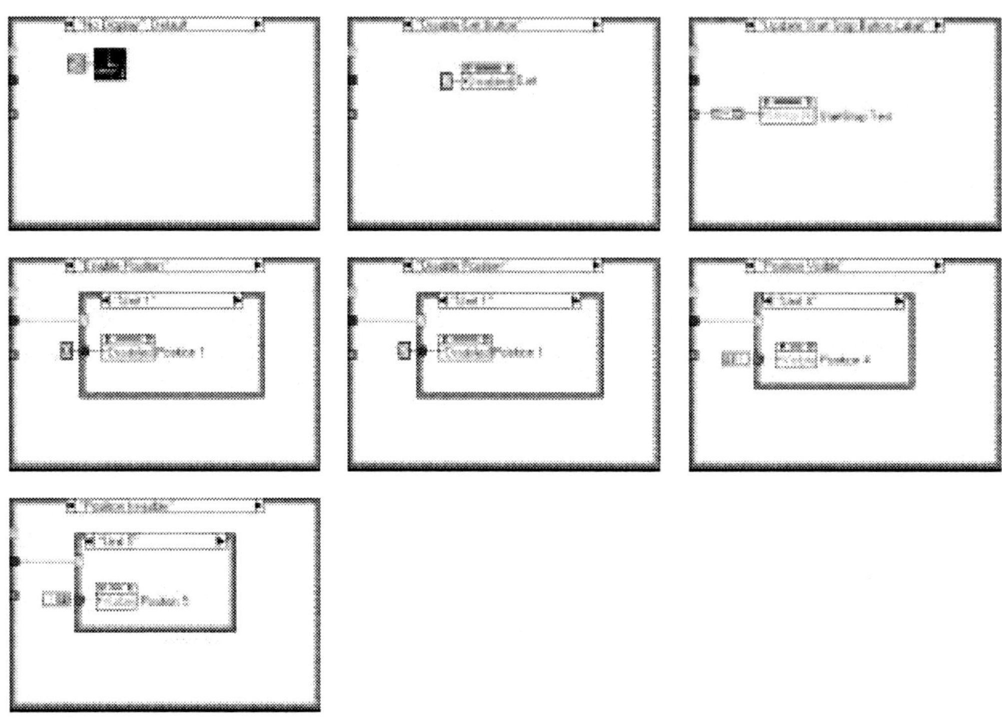

Figure 9.11
Control Display cases.

Figure 9.11 shows examples of the cases filled out. You will need a case for every action.

Now if you load a Test Stub VI called UI Display Stub.vi, and run the Widgetometer User Interface, you will be able to test the changes that have been made. The UI Display Stub.vi simply places messages on the message queue, and these are actioned by the User Interface. Now we are ready to give the prototype some life.

On starting, the program will settle into its first state: *Test Not Started*. When the start button is pressed, the *Enter Data* state is activated. After successfully entering data, the *Testing* state is started and the sequence of tests are run through. When the tests are done, the *Tests Complete* state is entered and any posttest actions finished.

A flexible approach is to produce a simple component that drives the program executive states. Figure 9.12 demonstrates how.

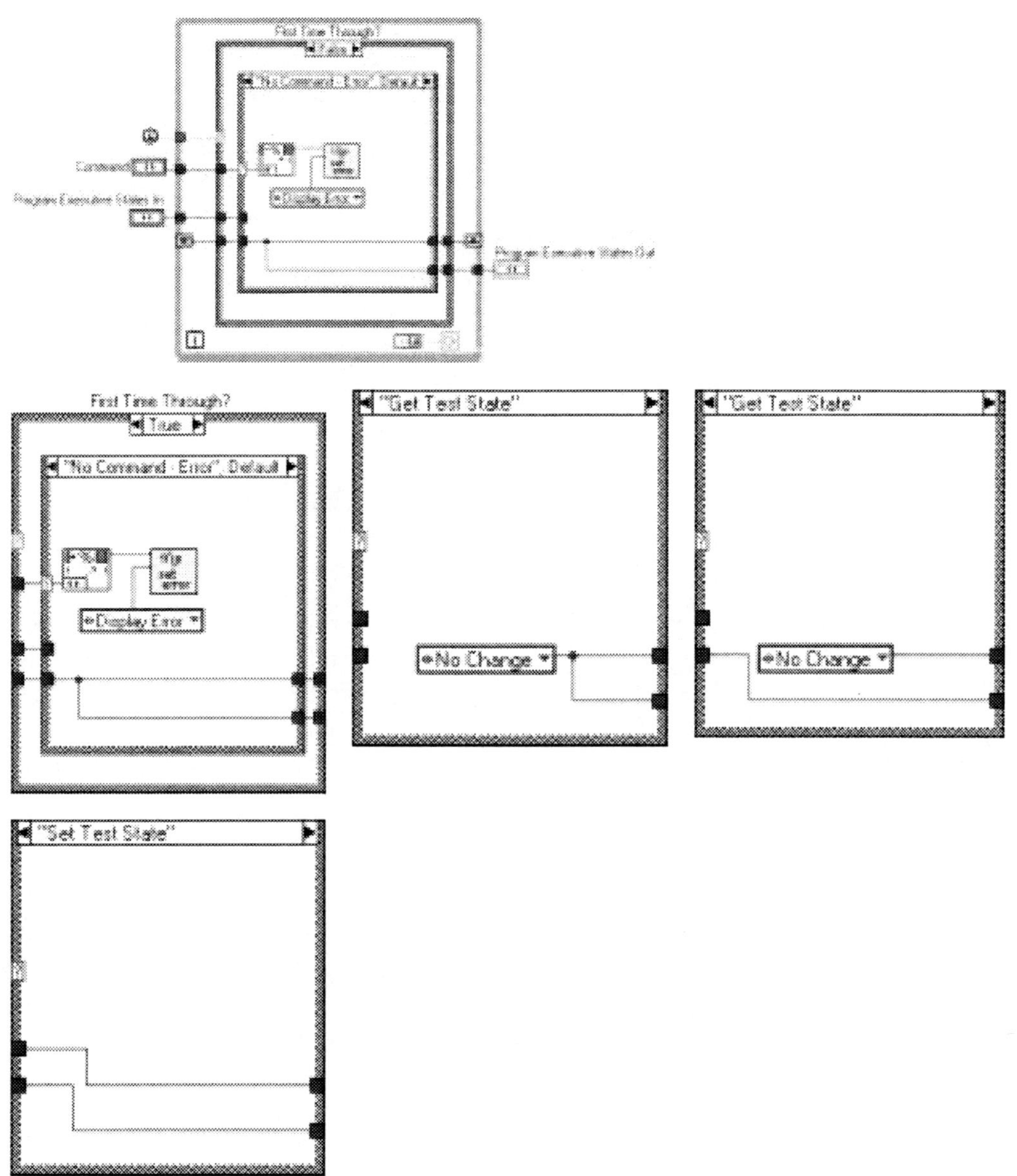

Figure 9.12
Program Executive State Controller.

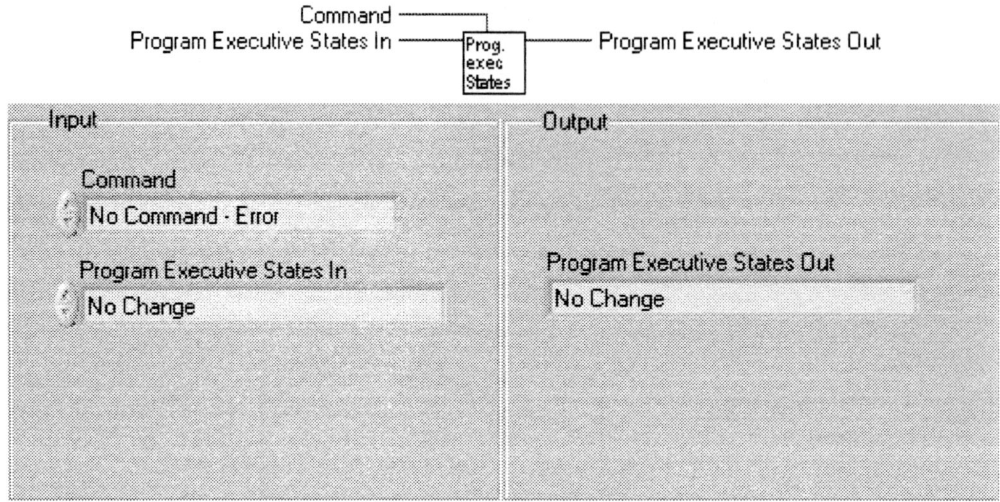

This component stores the required program executive state locally, by the "Set Test State" command, and returns it once on a "Get Test State" command. All we need to do now is have a loop checking the Test State and acting when it changes, and we have ourselves a nice, flexible program executive. We'll add the loop to the Widgetometer User Interface VI and put in some functionality, or the "Enter Data State." A simple Enter Data dialog box was created, and if you notice the Monitor Control Buttons loop on the diagram in Figure 9.13, you'll be able to see that when button 2 [Start Stop Test] is pressed, the Test State is set to Enter Data. When the Program Executive loop next iterates, it will retrieve that state and action it.

Inside the Enter Data dialog will be the code that decides what step to go to when the dialog exits. If it exits on a Cancel the next state should be to return to Test Not Started. If the dialog exits successfully then the next state would be Testing. In a similar fashion to decoupling the User Interface actions from the front panel, we've now decoupled the Program Executive actions. We now have the freedom to control the way the Test Sequence runs from nearly anywhere in the software hierarchy.

9.3   Build

Now we can proceed on our journey with the confidence that we have done everything possible to extract information from our customer. The customer

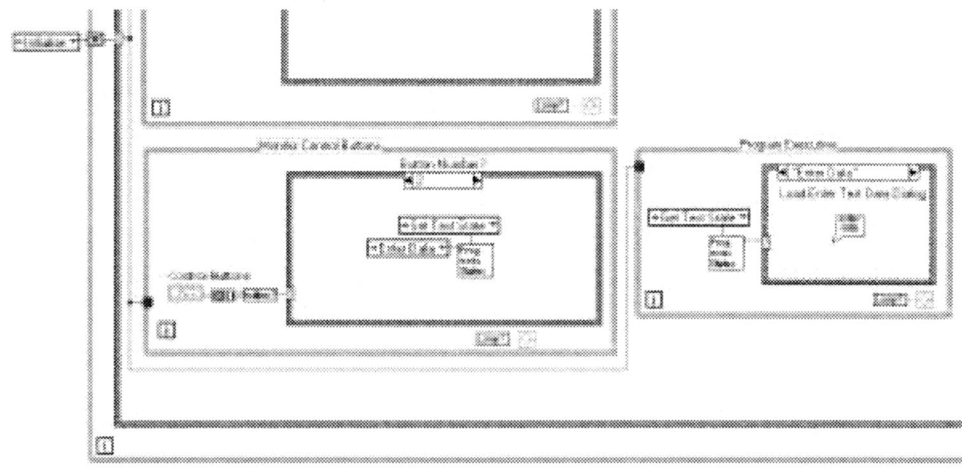

Figure 9.13
Program Executive Enter Data State.

in turn should have some confidence that the project about to be received will make him or her truly thankful.

So, onward.

9.3.1 Code and Fix

Progress was rapid, but now changes are having effects on the stability of the software, plus the customer design review threw up a few surprises. Anyway code-fix, code-fix, code-fix.

9.3.2 LCOD

Now we shall concentrate on adding functionality to the framework of components that we've identified, or that our pattern has identified for us. For the sake of simplicity, let's assume that the framework of components produced from the first principles agrees closely with the framework defined by the patterns.

9.3.3 Hardware

Looking closer at the hardware pattern, we've already defined the top-level commands for the component, so now we'll go deeper. Each hardware action is a combination of Control, Drive, and Read. There may be no need for all of them, but at least one will be required. For instance, a simple thermometer would only have the Read component, whereas a scanning thermometer would have a combination of Control and Read.

Here's the implementation for this project and a reminder of the actions for this component.

Hardware Actions

It transpires that the power is switched on to all the units at the start of the test and removed at the conclusion of testing. This means that two extra hardware actions are required to apply and remove power to the units.

So, here are highlights from the source code.

Action	Description
UnderRange	Throws or passes error
Initialize System	Sets up all the hardware components
Switch STDs In	Connects the standard units to the selected test position
Reset Switches	Clear connections
Read Unit Status	Read the selected unit's Status from the serial port
Read Unit Serial Number	Read the selected unit's Serial Number from the serial port
Read Unit Part Number	Read the selected unit's Part Number from the serial port
Read Unit Software Version	Read the selected unit's Software Version from the serial port
Read Chan1 Reading	Read the selected unit's reading for channel 1 from the serial port

Action	**Description**
Read Chan2 Reading	Read the selected unit's reading for channel 2 from the serial port
Get Chan1 An Out	Measure the selected unit's channel 1 analog output
Get Chan2 An Out	Measure the selected unit's channel 2 analog output
Get Chan1 Display Reading	Get the display reading for channel 1
Get Chan2 Display Reading	Get the display reading for channel 2
Power On	Apply power supply to units under test
Power Off	Remove power supply from units under test
OverRange	As UnderRange

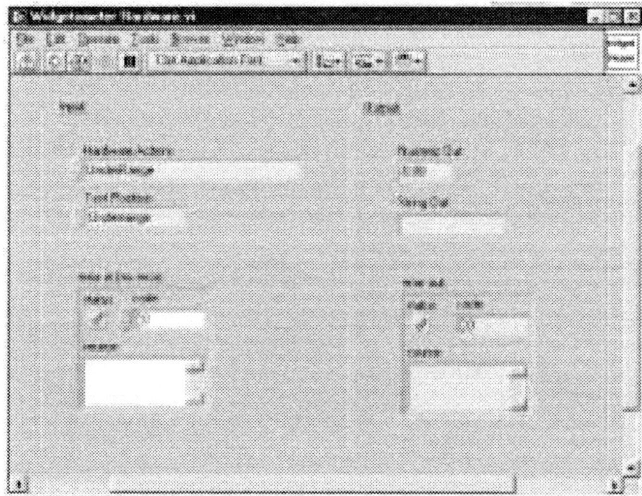

Figure 9.14
Hardware component front panel.

Looking at Figure 9.14 you can see that Test Position is a Strict Type Def. enumerated type that has already been defined for the prototype and is used throughout the software. Hardware Actions is a new Strict Type Def. enumerated type that has all the actions that were defined earlier.

Hardware Actions has been set to "required" as a precondition, and a standard wiring pattern has been defined, as shown in Figure 9.15 and 9.16. All components will follow this pattern where possible.

The block diagram looks like that shown in Figures 9.17. The first time through the hardware is initialized.

Figure 9.15
Widgetometer hardware icon.

Figure 9.16
Standard wiring pattern.

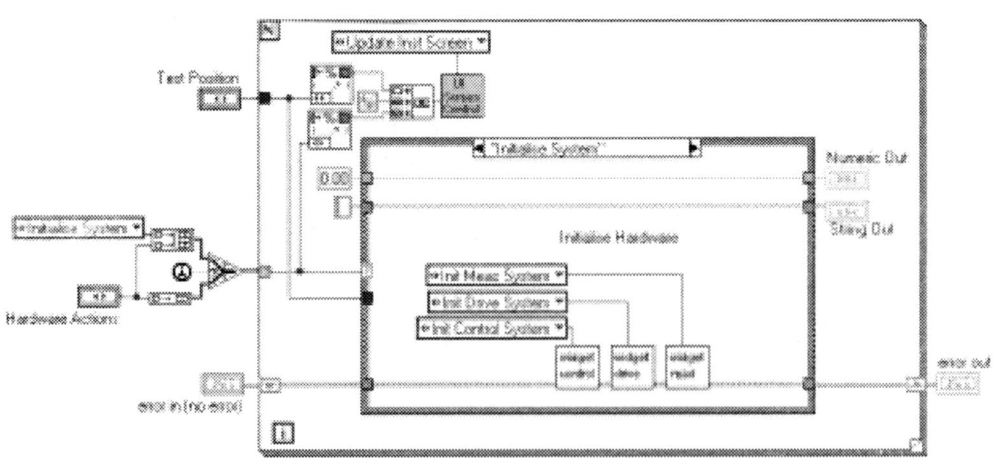

Figure 9.17
Hardware component block diagram.

The UI screen control is sent a message describing each action that has been taken by the hardware component. The hardware action is then implemented. The actions that talk to the serial ports are all pretty repetitive and involve sending the message and the unit position for that message to the read component. This will then do all the communicating and return a string or numeric value. The measure action is a bit more involved; it requires a bit of switching to select the unit under test.

Clearly demonstrated here is the self-documenting nature of LCOD—you can see exactly what each component is being asked to do.

Programming the Get Chan2 An Out action was simply a matter of duplicating the Get Chan1 An Out case and changing the command to the control component as shown in Figure 9.18. The Power On and Power Off commands use the drive component and simply set the power supply volts to 24 or 0 as shown in Figure 9.19.

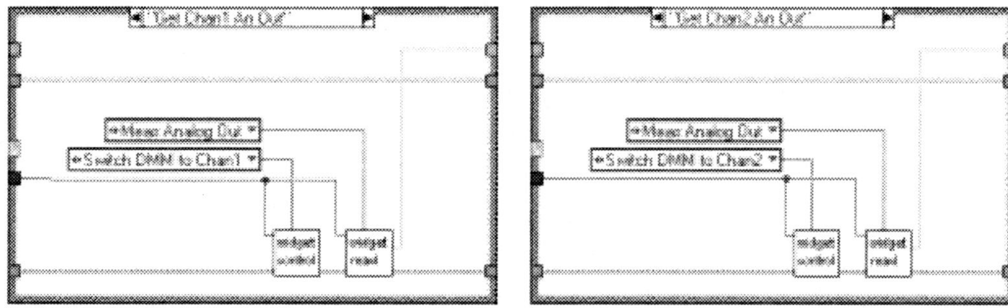

Figure 9.18
Get Chan 1,2 An Out.

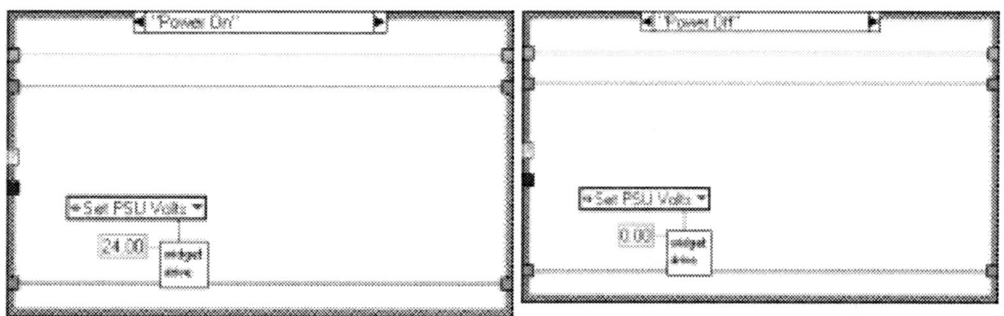

Figure 9.19
Power On/Off.

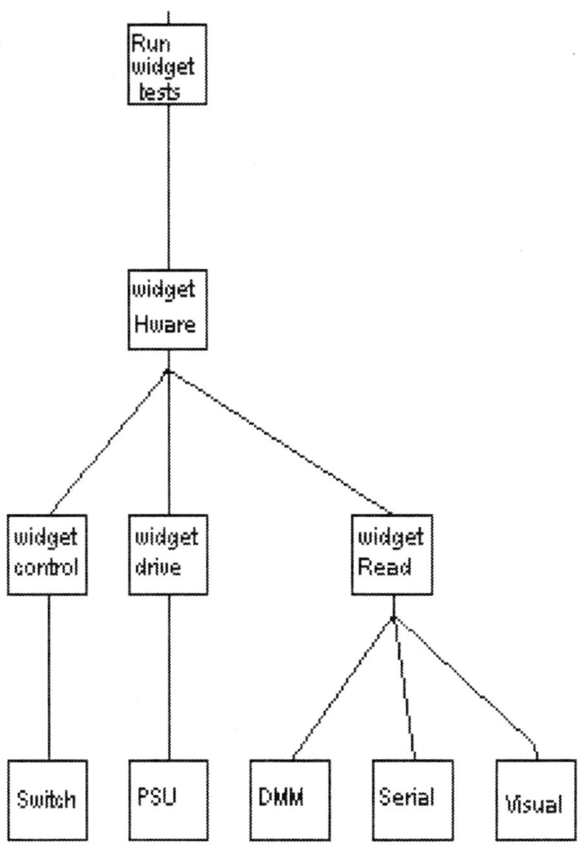

Figure 9.20
Widgetometer Control Drive Read pattern.

The hierarchy in Figure 9.20 shows the Control>>Drive>>Read pattern. Below the generic Switch, DMM, and PSU components would be the manufacturers or your own drivers. For continuity we tend to rewrite them as components as well.

Control Actions

It is envisioned that the measurement and signal lines are switched from the units under test to the computer and measurement equipment.

Action	Description
UnderRange	Throws or passes error
Init Control System	This initializes the control hardware, getting and setting identifiers and setting initial conditions
Reset Switches	Set all switches open
Switch Standards In	Switches the standards into the chosen position
Switch DMM to Chan1	Switches the DMM to channel 1 of the chosen position
Switch DMM to Chan2	Switches the DMM to channel 2 of the chosen position
OverRange	As UnderRange

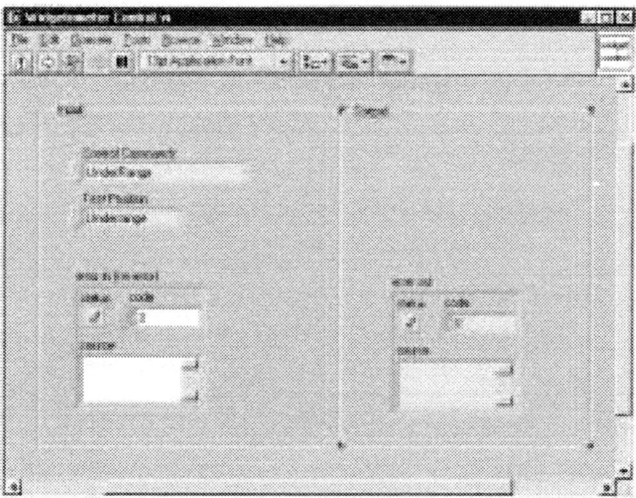

Figure 9.21
Control component
front panel.

Here's the implementation.

So regarding Figure 9.21 we will define Control Commands as a Strict Type Def. enumerated type and use Test Position as in the hardware component.

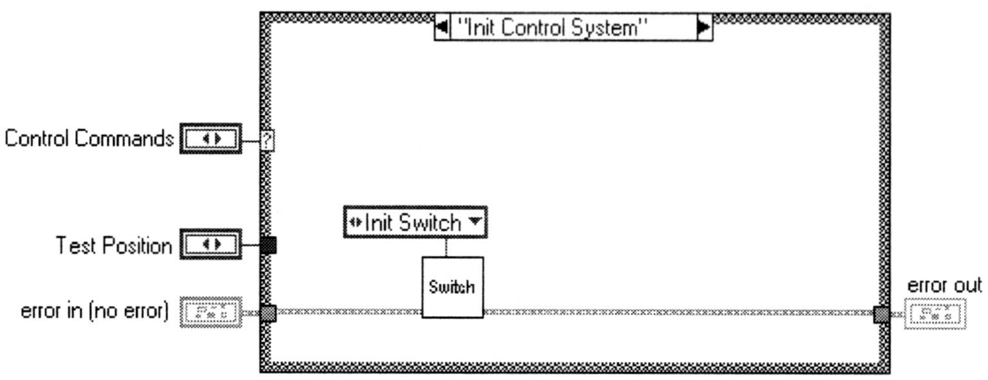

Figure 9.22
Control Component Initialize State.

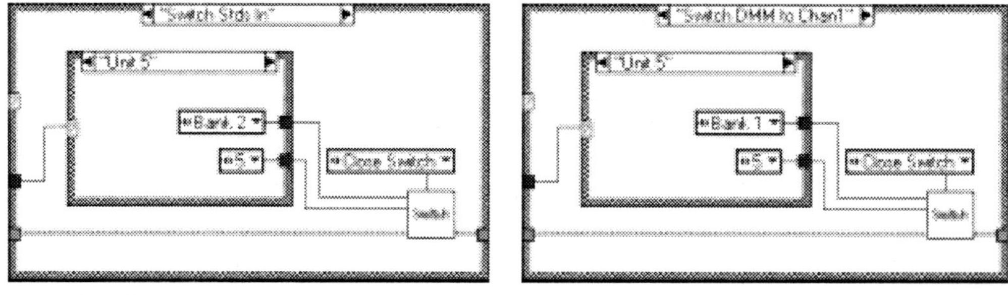

Figure 9.23
Control component switching commands.

As we go to lower levels of abstraction we can see the commands becoming more oriented to the hardware. This is demonstrated by the states in Figures 9.22 and 9.23 that are now less identifiable as part of a test system and more specific to switching systems.

Drive Actions

Drive is used for signal injection, PSU setting, and similar actions. For this project there is a requirement to switch on the power supply to the Widgetometer.

Action	Description
UnderRange	Throws or passes error
Init Drive System	This initializes the drive hardware
Set PSU Volts	Sets the power supply volts current
Get PSU Volts	Returns the power supply volts
Get PSU Current	Returns the power supply
Clear PSU	Sets the power supply to a safe level, 0 Vdc
OverRange	As UnderRange

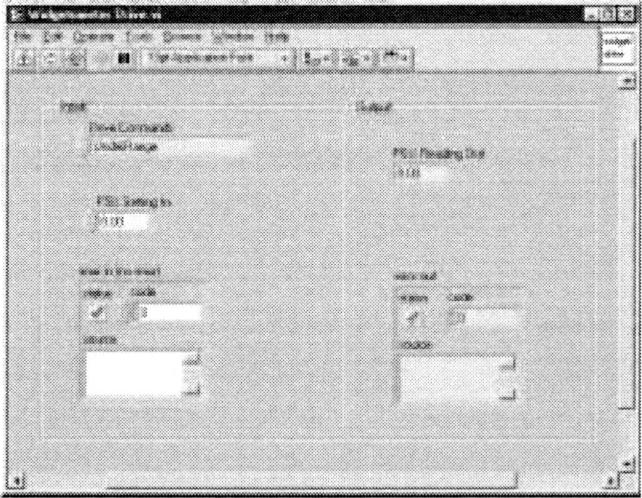

Figure 9.24
Drive component front panel.

The front panel, as shown in Figure 9.24, is a command Strict Type Def. enumerated type, with just the voltages in and results out.

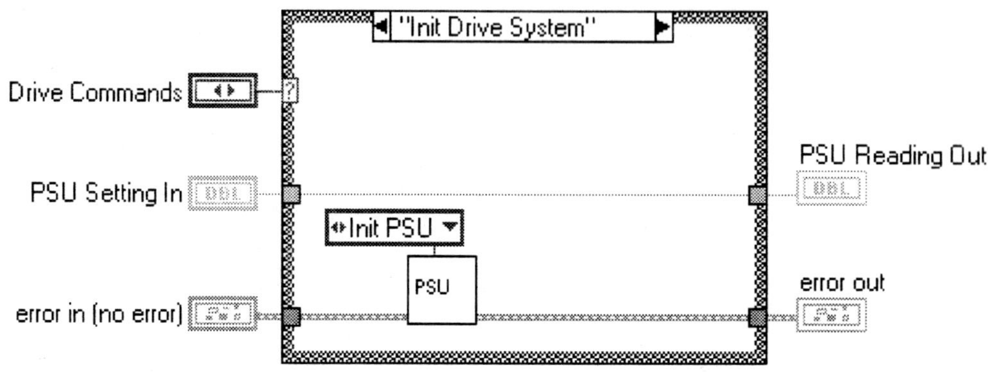

Figure 9.25
Initialize drive component.

The rest of the states are the same, as shown in Figure 9.25, with corresponding messages to the PSU component.

Read Actions

There will be a multimeter for analog input measurement. There will also be the commands to talk to the serial port. Finally, we'll put the visual inspection stuff in here as well.

Action	Description
UnderRange	Throws or passes error
Initialize Meas System	This initializes the measurement hardware
Read Unit Status	Read the selected unit's Status from the serial port
Read Unit Serial Number	Read the selected unit's Serial Number from the serial port
Read Unit Part Number	Read the selected unit's Part Number from the serial port
Read Unit Software Version	Read the selected unit's Software Version from the serial port

Action	Description
Read Chan1 Reading	Read the selected unit's channel 1 reading from the serial port
Read Chan2 Reading	Read the selected unit's channel 2 reading from the serial port
Get Chan1 Display Reading	Get the display reading for channel 1
Get Chan2 Display Reading	Get the display reading for channel 2
Measure Analog Out	Returns the dc volts from the DMM
OverRange	As UnderRange

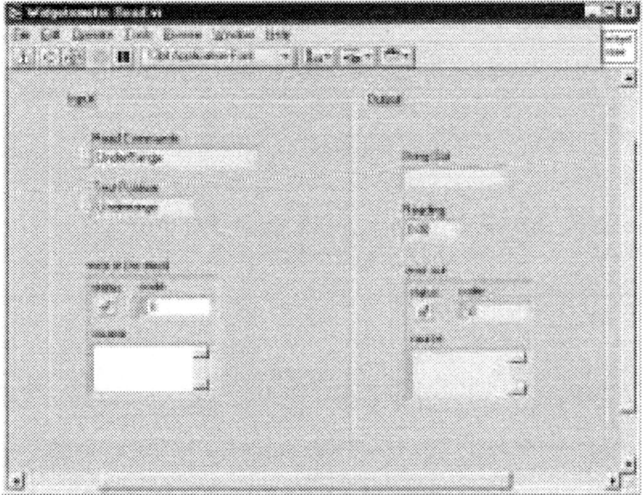

Figure 9.26
Read component front panel.

Let's quickly look at some of the implementation of these actions.

Figure 9.26 shows the Read component front panel, where once again we use the ubiquitous Test Position Strict Type Def. enumerated type, and as before we define a new Read Commands Strict Type Def. enumerated type with the commands defined earlier.

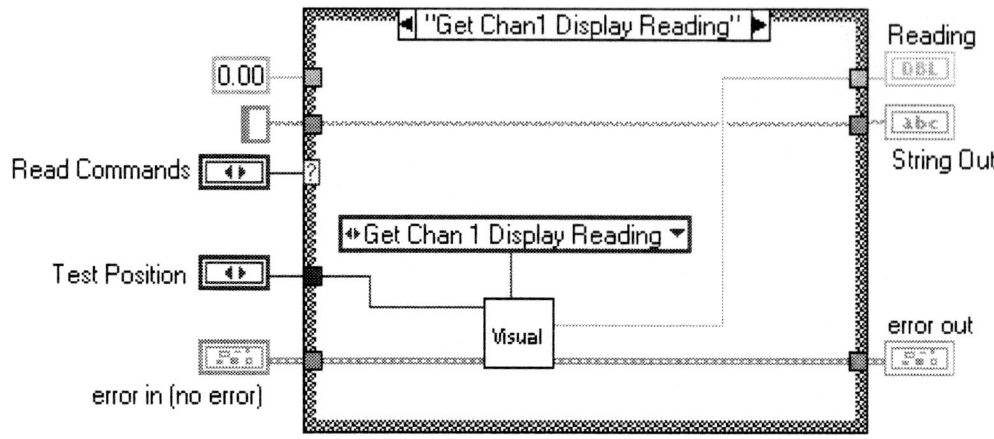

Figure 9.27
Visual component call.

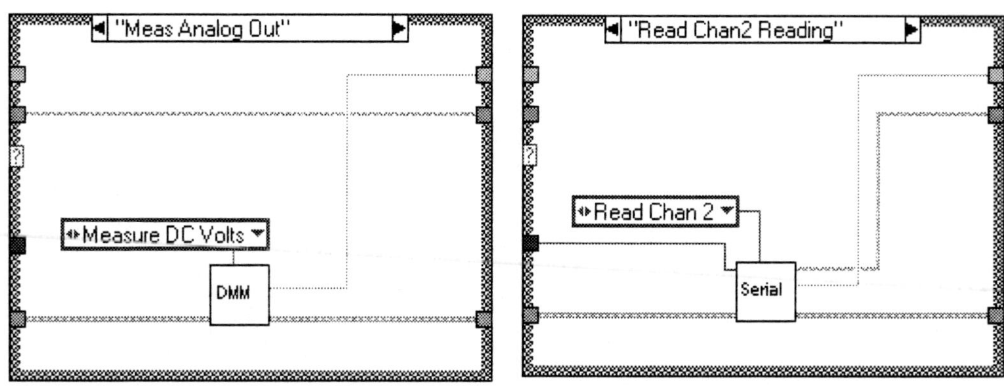

Figure 9.28
DMM and serial component calls.

Nothing too tricky about this. The read component merely calls the visual, serial, or DMM components and returns their responses as shown in Figures 9.27 and 9.28.

The only other action of note is the Initialize command, which is shown in Figure 9.29.

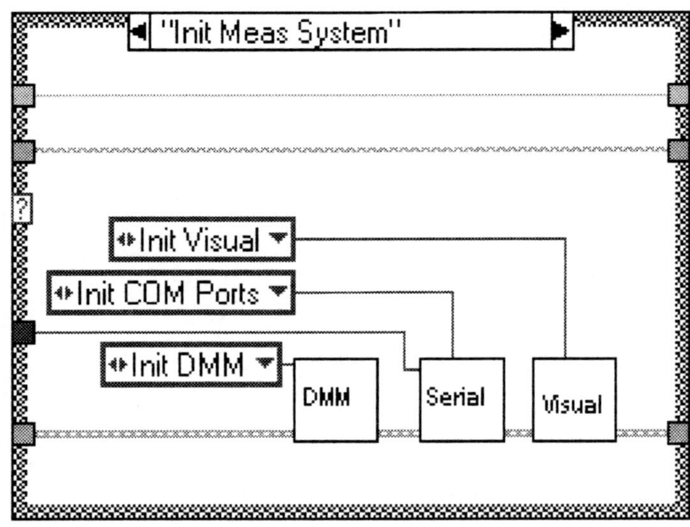

Figure 9.29
Read Initialize action.

Finally, you need a test executive, so that is discussed next.

The array of hardware actions drives the For Loop for each active test position. In reality, the results would go to a file or a database. The more observant reader will have looked at Figure 9.30 and noticed that the array "Tests" is an array of constants. Correct! Although they are dynamic Strict Type Defs., they are indeed constants and would be better placed in a file.

9.3.4 Detail Outside the Code

If you see a constant String or Number anywhere in your code, pull it out into a configuration file. We use a pattern for this that takes advantage of the flexibility of enumerated types.

Make a copy of the Prototype Software Directory on your hard drive and rename it Widgetometer Software Directory. Now you need to copy the configuration directories over to where the program resides. So copy the following directories to your Widgetometer Software Directory:

....\Patterns\Section Keyed Data Handling Pattern\1 Displays

....\Patterns\Section Keyed Data Handling Pattern\2 System

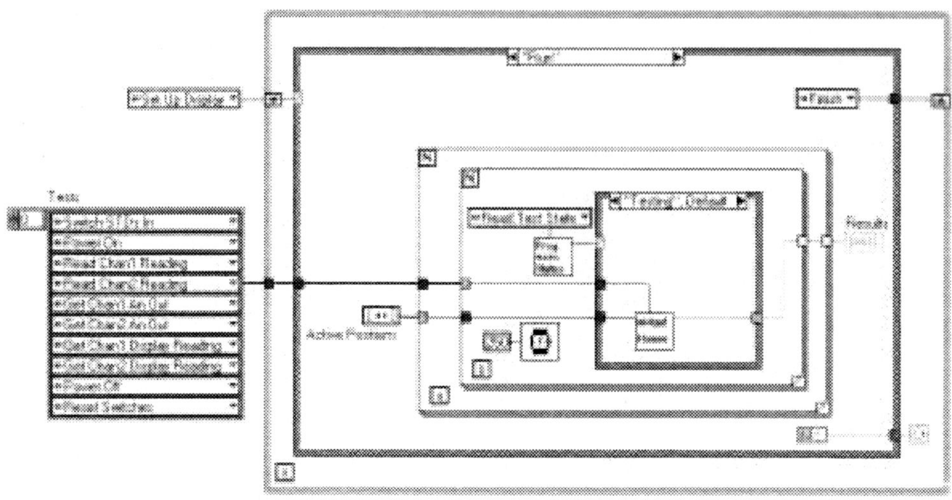

Figure 9.30
Widgetometer test executive.

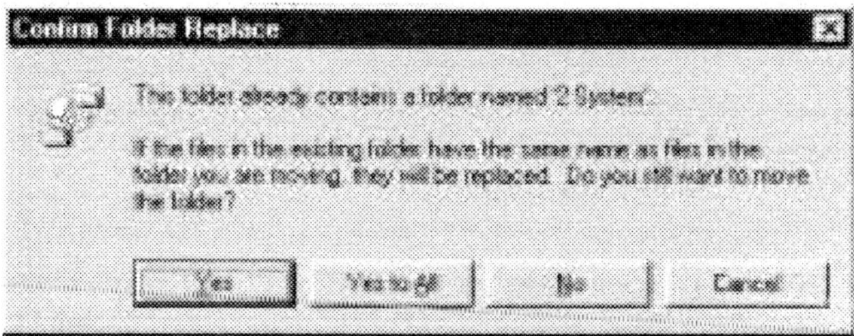

Figure 9.31
Confirm Folder Replace.

If this has been done correctly, you will find that the following message will be displayed as shown in Figure 9.31.
Select [Yes to <u>A</u>ll].

Rename the directories to suit as shown in Figure 9.32.

Open Widgetometer User Interface.vi and find any lost VIs. Widgetometer User Interface loses Widgetometer Enter Data.vi and 2 button dialog.vi. Then select File>>Save With Options>>changed VIs from the menu and save everything.

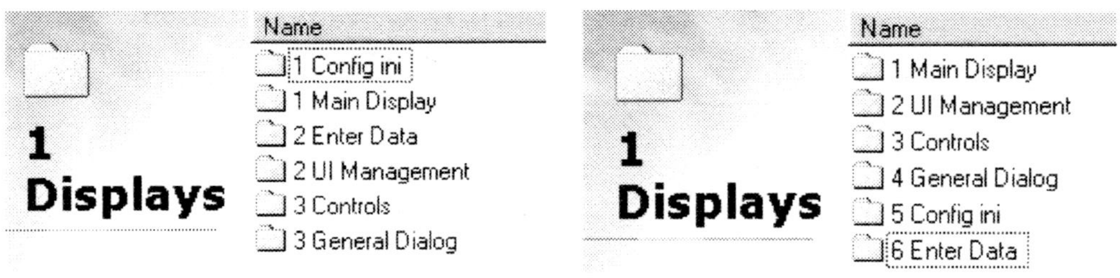

Figure 9.32
Rename directories.

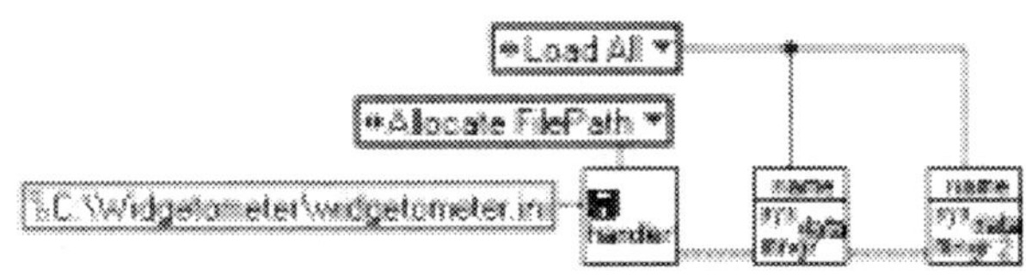

Figure 9.33
Add configuration to initialize VI.

Without closing Widgetometer User Interface.vi, open Config ini.vi and as before find any lost VIs or Controls. This is usually just Datamanager.ctl. Once again select File>>Save With Options>>changed VIs from the menu.

Rename Config ini.vi to Widgetometer Config ini.vi, and update the icon to reflect the new program.

Now we need to open the configuration file when the program starts and load all the configuration data. In Widgetometer User Interface.vi you will find UI Initialize.vi, open it up and put . . .\\2 System\2 Data Handling\1 File Management\1 File Handler\File Handler.vi into it, and put in the file path to the configuration file as shown in Figure 9.33.

Also put in the two data manager VIs. They can be found in . . .\\2 System\2 Data Handling\Config Data (or copied from the hierarchy diagram). Pop up on the command input and select "Load All".

Now drop Widgetometer Config.ini into the case for Button 3 in Widget-ometer User Interface.vi diagram (run state), as illustrated in Figure 9.34.

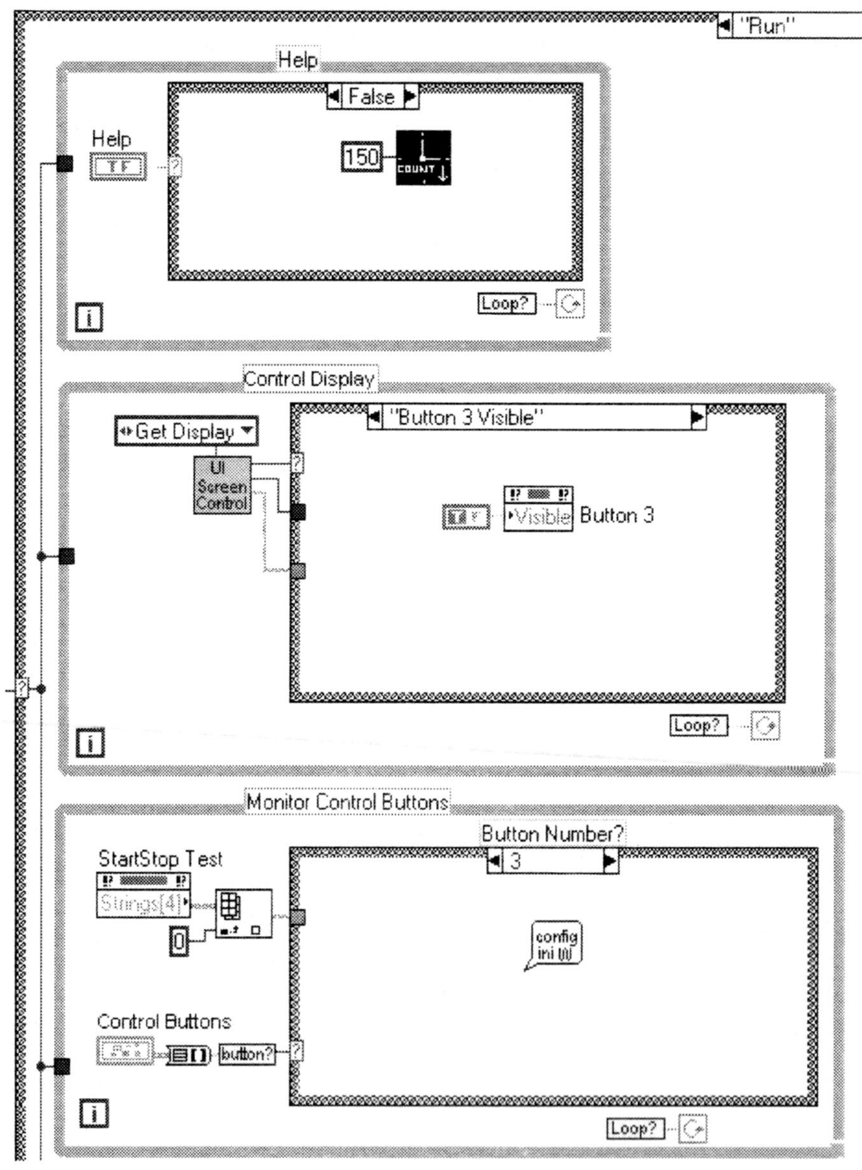

Figure 9.34
Make Config.ini dialog active.

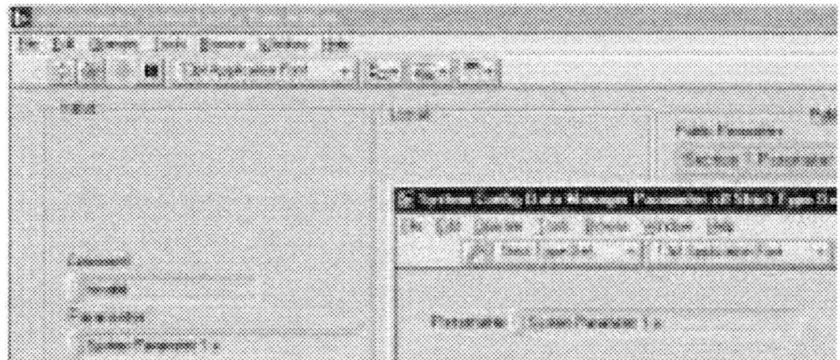

Figure 9.35
Customize parameter.

You can optionally change the names of the data manager files and customize the icons to your individual requirements. What you now have is the capability to create, load, and save constants in a reasonably secure location. Defining these constants is as simple as updating an enumerated type control.

Let's do a couple as an example. Let's remove the example parameters and add a new parameter called System Name. We've changed the data manager names to Widgetometer System Config Data Man.vi and Widgetometer System Config Data Man2.vi.

Double-click on the parameter control on Widgetometer System Config Data Man.vi front panel to bring up the control editor (if you haven't got this option set you'll have to open the control in longhand by highlighting the control and selecting customize from the menus). Figure 9.35 shows the resulting screen.

Remove the example parameters and replace the last one (you won't be able to remove it) with System Name.

To run the new set up you will have to ensure that the directory for the Config.ini file exists (as pointed to in UI Initialize.vi). The ini file will be automatically created when the program is first run.

Now run Widgetometer User Interface.vi and select Button 3 (after it moans about there being no file). The Config.ini dialog should be displayed. You will notice that there is now a key called System Name.

Other constants we could pull out of the code at this stage are:

Power Supply On Volts
Power Supply Off Volts

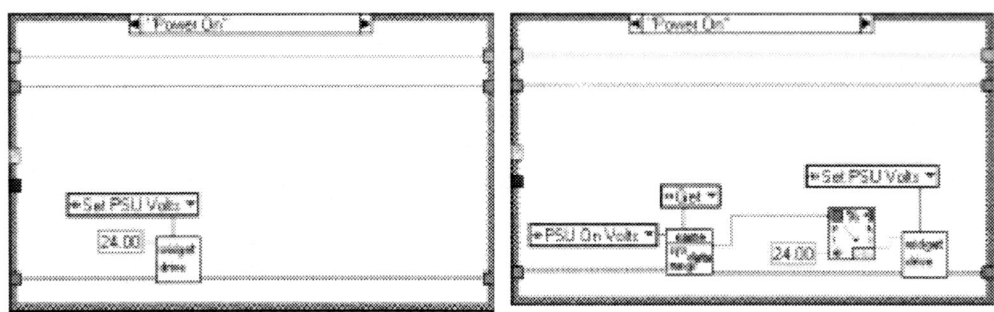

Figure 9.36
Taking detail out of hardware VI.

Serial Port Settings (baud rate, handshaking protocols and data, and start and stop bit sizes)

All you do then is drop the Widgetometer System Config Data Man.vi into the relevant hardware VI and wire it as shown in Figure 9.36.

Finally, the array of Strict Type Def. enumerated types used to drive the Test Executive should also be generated from the configuration file.

9.3.5 Error Handling

In keeping with the components within components theme we have created an error handling component. This component can be set to a mode where it complains loudly (throws dialogs) or logs errors discretely. In development mode you want it to complain loudly, and in run mode you want it to store the errors in a file. This component is called Error Control.vi and comes with both the User Interface Pattern and the Config.ini pattern.

Information hiding makes the error handling decision-making process a little simpler. Lower-level private components can pass errors up to their peer components. These errors can then be handled by the top-level components. The following diagram, as shown in Figure 9.37, illustrates the process.

The program executive could also have an error state that handles the errors and warnings in an elegant manner.

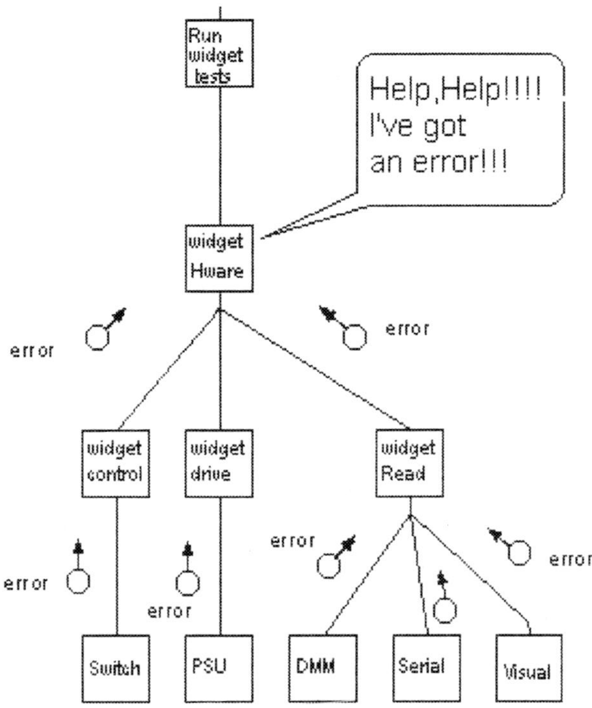

Figure 9.37
Widgetometer hardware error handling.

9.3.6 State Machines

The program executive was discussed earlier, but what are the advantages of doing it like this? It is actually quite easy to list the different states of a system. From a hardware perspective, most test systems have very few states. These states are then defined by the hardware component. So using state machines has simplified the definition of the problem.

The program executive is usually just as simple to define. The ones in Figure 9.38 describe a common pattern.

Another common state machine is the Test Executive. The one used in this system takes advantage of the fact that the hardware component models all the states that the test system has. It takes an array of these states and sticks

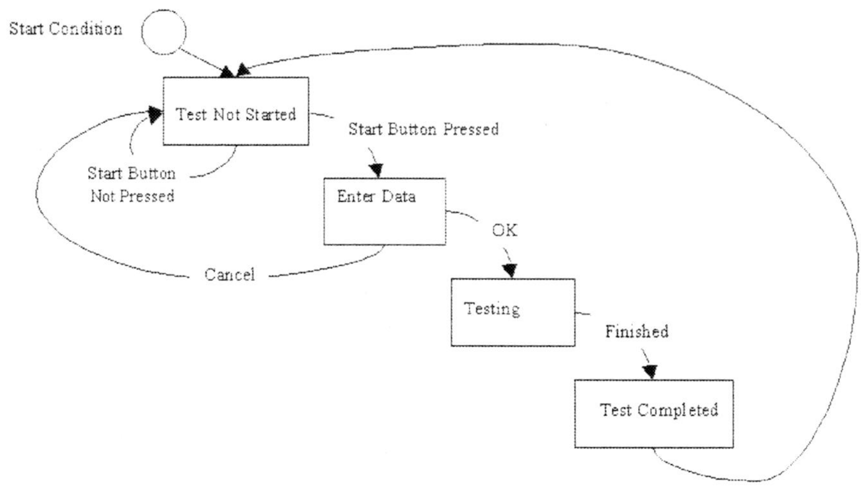

Figure 9.38
Widgetometer program executive.

them into a For Loop, creating a sequence of states that defines the test. For extra flexibility this array of constants should be pulled out of the code and stored in a file.

The state machine structure also helps when changing code. In the Widgetometer case if we wanted to define and add a completely new test, we would simply update the hardware component and add the new hardware actions to the array in the Widgetometer test executive.

9.3.7 Reuse

What have we reused and how much effort has it saved us?

Patterns

Patterns allow us to reuse design and also enable us to reuse more of our code.
For this project the following patterns were used:

UI Controller-Message Queue Pattern

Section Keyed Data Handling Pattern

Control>> Drive>>Read Pattern

VIs With Little or No Modification

It's all right, we're not going to list them all. A quick count tells us that without including any of the hardware, we have reused nearly 70% of the code in some shape or form.

Reuse can sometimes have implications on the design, since you can end up changing your design so that you can reuse existing code. This shouldn't be the case with LCOD because the patterns are purely a way of emulating the patterns observed in all systems when you start using components in your code. Strict Type Defs. then allow you to modify the pattern to suit your design.

Also, what new code can you reuse in future systems? For a start, you can pluck out the lower-level hardware VIs.

9.3.8 Style

If you are in a multideveloper environment you will find the following checklists useful for broadcasting the standards expected for a job. They are also quite good as reminders, even if you're working on your own. We like checklists!

DEREK'S SOFTWARE CO.

Description	Widgetometer Test System Style Checklist
Document Number	SCL_WATE10001
Issue	1.00

Description	Status
General Layout Standards	
Are diagrams as compact as possible while still having a visible flow?	
Is the block diagram crammed?	
Any floating structures?	
Does the block diagram fit on one page?	
Wiring Standards	
Left to right data flow	

(continued)

Description	Status
Wires should only overlap if there is no alternative	
Wires should be as straight as possible	
No data into structures from the top or bottom	
No wires routed through an icon to a terminal on the other side of the icon	
No wires routed underneath structures or icons	
Inputs and outputs aligned	
Excess wires such as loops deleted	
Labeling Standards	
Every frame in a Sequence, Case, or Loop structure has a descriptive comment. These should be indented if there are embedded structures.	
Comments for wiring should have a white background with no border.	
Boolean cases should be labeled to document the case states.	
While Loops and For Loops should be labeled to document their purpose.	
Shift registers should be labeled with their data contents.	
Font Style and Size	
Control and Indicator labels Arial, Normal, Size 13	
Public Front Panel Standards	
Strict Type Definitions used for Enumerated Commands and Attributes	
Icon filled in	
A user description for each control or indicator	
Private Front Panel Standards	
Connector Controls on the left, Connector Indicators on the right, Local Indicators and Controls in the middle or on the bottom	
Align and evenly space Controls and Indicators	

Description	Status
Controls and Indicators in the same order as they are connected	
Frame the relevant areas Input, Output, and Local where applicable	
Use the recessed frame from the decorations menu and label or the dialog frame	
Use the brush with the foreground set to transparent to tidy up the label	

9.4 Uh-oh We've Been Given the Wrong Directions

On rare occasions (like every project we've ever worked on!) the customer changes his or her mind, or wants extra functionality. This happens because the customer doesn't fully comprehend the system until seeing it and using it. This will result in a change in requirements. You can regard this as either an opportunity to shine or an opportunity to have a cardiac arrest.

The customer doesn't like the concept of entering data by dialog boxes, so he or she has decided to automate the test by reading the BCD inputs to the LCDs, provided by a test port. There are four digits of four bits for each card. With some digital multiplexing we can switch between displays on the unit under test and then between units. This can be done with a cheap Digital I/O card and a little bit of circuitry.

```
Port 1 Bits 1–4    - Digit 1 (least significant digit)
Port 1 Bits 5–8    - Digit 2
Port 2 Bits 1–4    - Digit 3
Port 2 Bits 5–8    - Digit 4 (most significant digit)
Port 3 Bit 1       - Display 1 & 2 Switch
Port 3 Bits 2–5    - Unit Selector
```

What do we have to change and where? Well, we know that all the tests are in the hardware component, so we can start there. Note how strong cohesion cuts down on all the hunting around for affected VIs.

So, at the hardware component we don't need to change the commands, and the public interface remains the same. As far as any VIs are concerned nothing has changed.

The actions affected are:

> Get Chan1 Display Reading
>
> Get Chan2 Display Reading

The changes are shown in Figures 9.39 and 9.40.

Three new actions have been added to the control component to set the display to be read (1 or 2), and to switch in the digital multiplexer for the required test position. We've also improved the read component slightly by combining the "Get Display Reading" commands into one generic command.

Figure 9.41 shows the new control component additions.

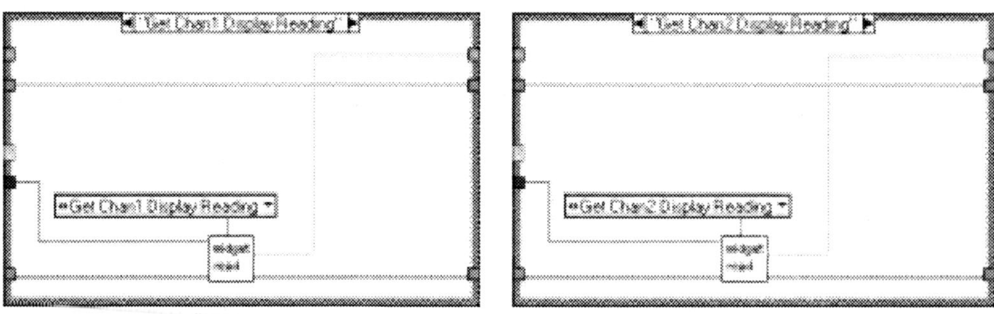

Figure 9.39
Hardware before changes.

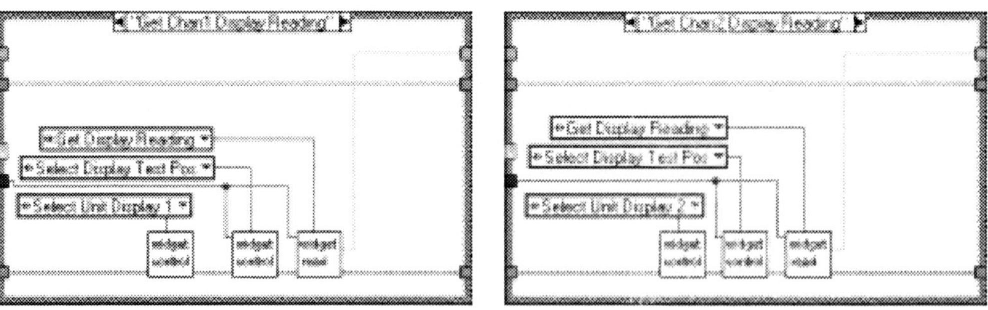

Figure 9.40
Hardware after changes.

And you'll notice that a new DIO component has been created. Figure 9.42 shows what it looks like inside.

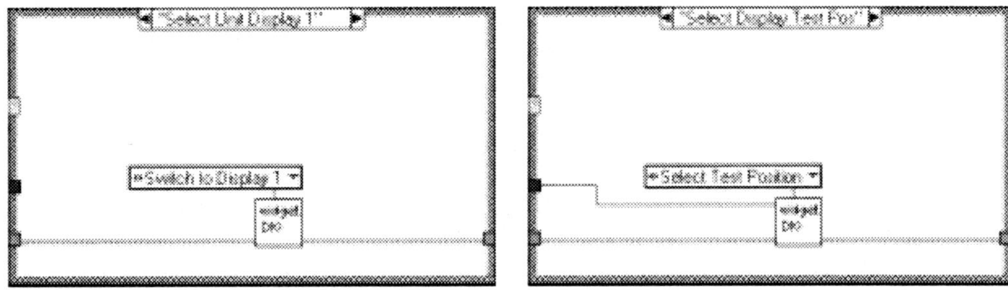

Figure 9.41
New control actions.

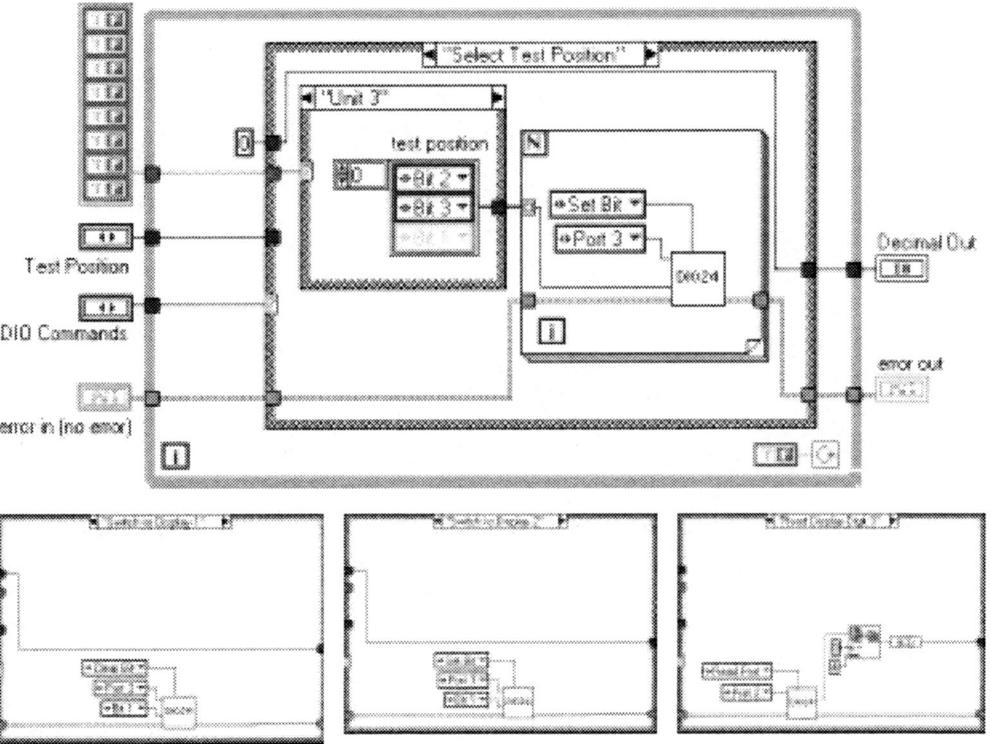

Figure 9.42
DIO component.

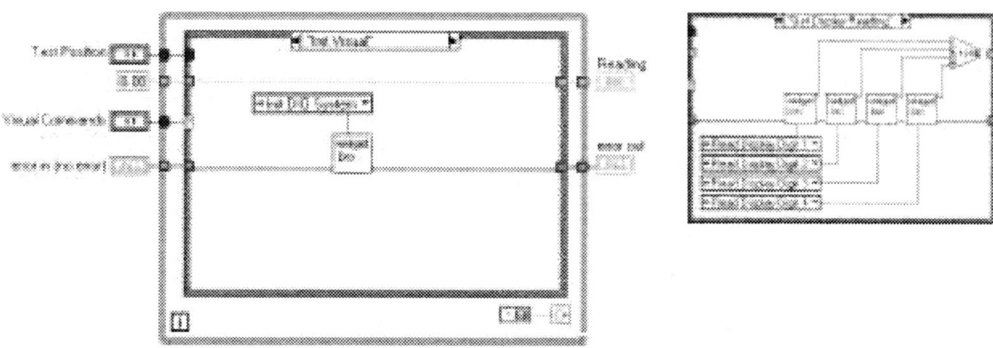

Figure 9.43
Visual component.

The read component changes consist of deleting the "Get Chan2 Display Reading" case and its corresponding command from the Read Commands Strict Type Def., and changing the "Get Chan1 Display Reading" to just "Get Display Reading".

The visual component is then modified to include the new DIO code for reading the BCD inputs to the displays. These changes are shown in Figure 9.43. For convenience we've committed a bit of naughtiness here; we're going to use the DIO component in the read component as well. Oh well, rules are meant to be broken. We should move the DIO component from its private directory to a public directory though to indicate that it is now being shared.

This clearly demonstrates the improved maintainability gained by incorporating information hiding, strong cohesion, and loose coupling. The result is that the robustness of the system has not been compromised by a reasonably sized design change. For the sake of brevity we'll only discuss the impact of the next change.

The customer would like a history of the actions of each test in the sequence displayed on the front panel. What we could do is shrink the current status indicator and add another new indicator called History. If you set the text small and use the Courier font it will look somewhat like a printout.

To implement it you will need to do the following, as shown in Figure 9.44.

Add two new commands to the UI controller commands and settings controls: "Clear History" and "Add to History". You do this by adding the commands to the end of the respective enumerated types. Duplicate the cases in the UI Controller component and implement the new commands.

In the Widgetometer User Interface VI add the new cases to the case structure in the Control Displays loop and put in the string manipulation required. Figure 9.45 shows an example.

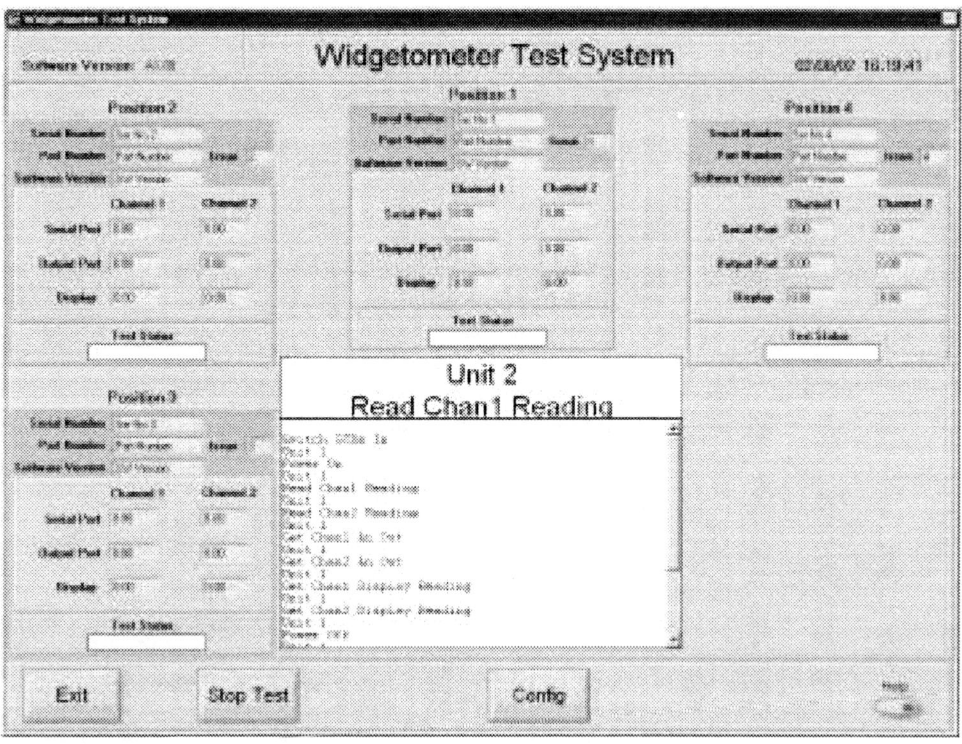

Figure 9.44
New History Display.

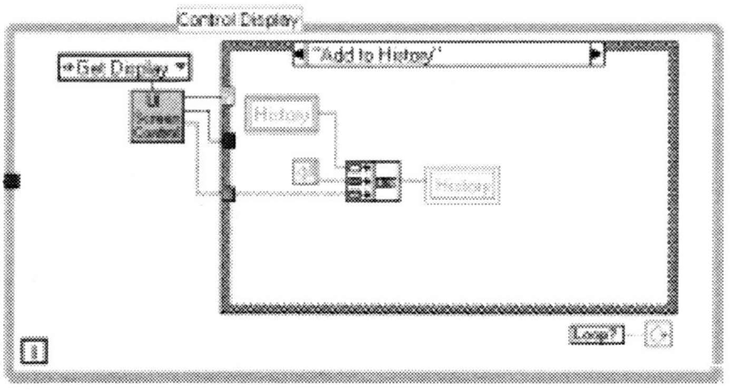

Figure 9.45
Add History Display action.

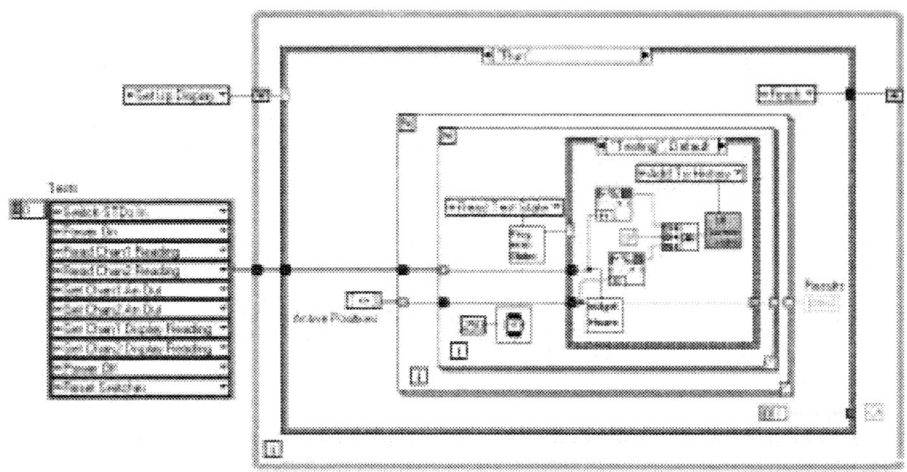

Figure 9.46
Update History in test executive.

You can then either send all calls to the hardware component to this display, or format something nice in the test executive and send that. The test executive option is demonstrated in Figure 9.46.

Review Sections 9.2.4 and 6.2 for more information on the ease of updating displays by implementing LCOD message-sending techniques.

9.5   Conclusions

So, the conclusion to our journey will be a happy ending. We provide the widget test software and the customer is blown away with the sheer brilliance of our software. We test the application against the test plan and it passes with flying colors. We are heralded as the masters of LabVIEW and a ticker tape parade is held in our honor . . . complete fantasy!

That would make our example project the worst fairy tale of all. Undoubtedly, problems will occur. Our example project could twist and turn with remarkable speed. The problems that could manifest themselves include:

- The market changes and new functionality *has* to be added.
- The people participating in the project change, bringing a whole new set of requirements and perceptions.

- Problems in third-party software rear their ugly heads and cause havoc with the system performance.
- The customer realizes that very late into the project, he or she has missed a fundamental test that would render the equipment useless without it.

How do we cope with all the possible problems and unknowns?

Hopefully, everything that we have discussed has shed light on what will give you a decent chance for success. LCOD with basic software engineering practices has enabled us to approach any project armed with the tools that will get the job done. Concentrating on loose coupling and strong cohesion allows the system to be defined simply. Information hiding allows the system to be designed with available information. These simple design strategies have long stood the test of time, and will remain doing so.

We know that using these techniques will help tame the complexity monster that all but the simplest projects can turn into. We know this because we have used them.

The techniques in this book are purely good design practices; they are not the latest "silver bullet." Everything we have presented is based on our experiences with LabVIEW and commercial projects. The techniques fit well with LabVIEW and data flow programming, and do not involve learning new abstract concepts and paradigms. As we stated very early on, none of the material in this book is new. We are constantly trying new things, keeping what works and discarding what doesn't.

We encourage you to do the same.

Glossary

ActiveX   A loosely defined set of technologies developed by Microsoft—we're talking about ActiveX components.

ADA   A programming language.

ATE (Automatic Test Equipment)   Tests the products under computer control.

Auto-Indexed   If you put an array into a For Loop and don't wire the number of increments, it will iterate every element in the array, passing that element data into the Loop.

Baud   The rate at which you can get your serial port to talk.

BCD (Binary Coded Decimal)   A numbering format for describing decimal numbers in 4 bits.

Bespoke Software   Software that is designed from scratch to the customer's requirements.

C++   A programming language that is object oriented.

CASE Tools (Computer Aided Software Engineering Tools)   CASE Tools help to automate the software engineering process by linking the design, documentation and management.

Cluster   A LabVIEW variable holder.

Coercion Dot   LabVIEW puts it on the diagram when converting one representation number to another.

Cohesion   A software design term, see text.

Component Interaction Diagram   A basic component-based data flow diagram.

Coupling   A software design term, see text.

ctl   Suffix of a saved control.

CVI   National Instruments' C-based software.

DAQ   National Instruments and others use it as a generic term for data acquisition equipment. Also, National Instruments' own line of internal PC cards for data acquisition.

Data Flow Paradigm   Data flows in on the inputs and out on the outputs and is processed in between. A program is several of these structures linked together by their inputs and outputs.

Decoupling   Removing the connection.

Delphi   Borland's "Visual" Pascal.

DFD (Data Flow Diagram)   The notation used for structured analysis.

Digital Multiplexing   Allows you to monitor many ports by the use of enabling signals. You make certain ports live by setting a clock or input.

DIO (Digital Input and Output)   What you use to get your 1s and 0s in and out of the computer. Used for driving relays, detecting switches, interfacing and so on.

DMM (Digital Multimeter)   A piece of measurement equipment which can measure Volts, current, resistance and more.

DSP (Digital Signal Processing)   Manipulation of analogue signals that have been converted into a digital format.

Encapsulation   A software design term, see text.

Enumerated Type   An unsigned integer in the form of a list.

G   LabVIEW.

GPIB (General Purpose Interface Bus)   Also referred to by its original name, HPIB, and IEEE-488, its standard. The GPIB was invented by Hewlett-Packard at the end of the 1960s.

GUI (Graphical User Interface)   The mouse driven user interface we've become so used to. Rather than the text based interfaces of the past.

HPBasic   A programming language that is Hewlett-Packard's test language for their 680×0-based workstations. It also comes in PC and UNIX flavors.

I/O (Input/Output)   Input is the data going in, output is the data going out.

ICs (Integrated Circuits) A small semiconductor device.

Information Hiding A software design term, see text.

Java A programming language that is Sun Microsystems' tidied up version of C++, with the pointers removed.

LabVIEW (Laboratory Virtual Instrument Engineering Workbench) A programming language.

LCOD (LabVIEW Component Oriented Design) A method by designing software using LabVIEW.

LED (Light Emitting Diode) Pass current through it and it will light. Many LabVIEW controls emulate these.

Merge VIs A way to reuse block diagrams.

Methodology A way of working that provides the framework to accomplish a particular function.

Modularity Breaking a problem into smaller "modules" makes the design process more intellectually manageable. Modularity is the measure of the number of modules used in a software design. For module think subVI or subroutine.

Multithreaded A single program can have several operations running in the processor at any one time.

MUX (Multiplexer) A method of splitting a single signal into many. We are referring to a type of switching system.

NI (National Instruments) The company that created LabVIEW. They want to sell you lots of hi-tech goodies.

μV A microvolt.

Noun A word used to describe.

Object Oriented A software design technique where rather than concentrate on functionality the designer searches for objects, their attributes, and the relationships between them.

OOA (Object Oriented Analysis) Understanding the problem using Object Oriented techniques.

OOD (Object Oriented Design) Translating the results from the analysis stage into a software blueprint. One advantage being that the notations from the analysis stage are the same.

OOP (Object Oriented Programming) Converting the design into the finished application using Object Oriented methods.

Paradigm A set of assumptions, concepts, or values.

Pascal A programming language.

Polymorphic   Can adapt to many shapes and data types.

Primitives   These are the lowest level base functions in a language, as provided by the language vendor. Nearly everything is written using them.

Problem Domain   Everything to do with the problem for which your software will be the solution.

PSU (Power Supply Unit)

QuickBasic   A programming language that is an older version of BASIC. It was the precursor to Visual Basic, without the visual part.

Ring-fence   When you drag a load of code with your cursor and turn it into a subVI by using Edit>>Create subVI.

RS232   A serial communications standard.

Scope   An abbreviation for oscilloscope.

Shift Registers   Used in While Loop and For Loop structures to pass data from one iteration to another. You can also use them to hold data in a subVI.

sig.gen.   An abbreviation for signal generator.

SmallTalk   A real OO language.

Source Code   The stuff in your block diagram.

State Transition Diagram   A diagram for mapping the states and transitions of state machines.

Strict Type Definition   If you set your control to this, any instances of it, including constants, will be updated when you update the original. The Strict part ensures that all the instances are physically similar throughout your software.

Structured Software Design   A design methodology that is the basis of LabVIEW.

Syntactical Errors   You typed it wrong.

TCP/IP (Transmission Control Protocol/Internet Protocol)   Your network or Internet probably uses it to communicate between computers.

Test Stub   A VI designed purely to test another VI.

Transducer   Converts one form of energy to another.

Tunnel   When data goes into a structure it makes a tunnel (little blob).

Turbo Pascal   A programming language—Borland's version of Pascal.

UCLA Pascal   A programming language.

UI (User Interface)   The users method of interacting with a system.

UML (Unified Modeling Language)   A combination of older Object Oriented modeling languages. Used to describe systems.

VB (Visual Basic)   A programming language—Microsoft's rapid development language.

Verb   A doing word.

VI (Virtual Instrument)   A LabVIEW program, so called because they originally imitated real instruments on your computer.

Wrapper VI / Component   This is a VI you put around another VI to make it more abstract and to simplify its interface.

Index

Advanced Topics in LabWindows/CVI

Shahid F. Khalid · ©2002, Paper with CD-ROM, 464 pp., 0-13-089229-7

The first easy to read, user-friendly book on advanced LabWindows/CVI features, with helpful hints, sample projects, and fully compiled and tested code samples.

Virtual Bio-Instrumentation: Biomedical, Clinical, and Healthcare Applications in LabVIEW

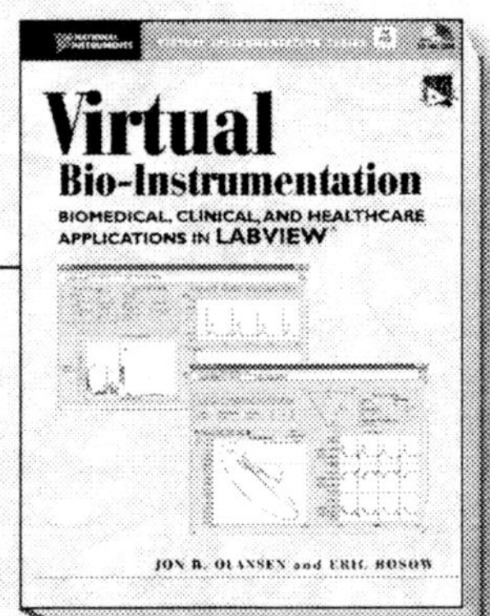

Jon B. Olansen / Eric Rosow · ©2002, Paper with CD-ROM, 603 pp., 0-13-065216-4

A first! - This book opens the boundless potential of Virtual Instrumentation (VI) into the wide variety of disciplines that exist within the biomedical and healthcare arenas.

LabVIEW for Data Acquisition

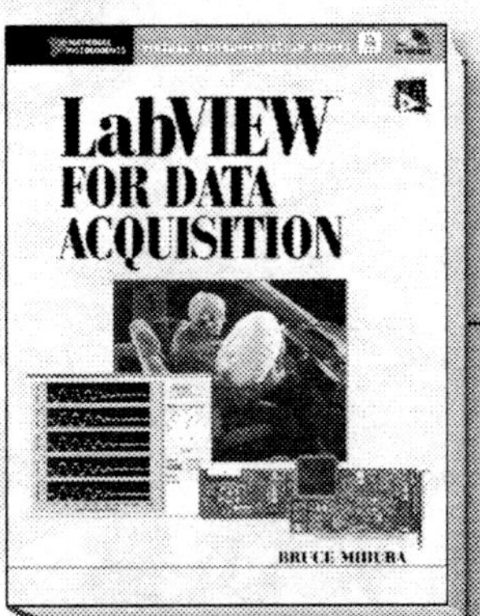

Bruce Mihura · ©2001, Paper with CD-ROM, 462 pp., 0-13-015362-1

The practical, succinct LabVIEW data acquisition tutorial for every professional.

Internet Applications in LabVIEW

Jeffrey Travis · ©2000, Paper, 624 pp., 0-13-014144-5

Learn how to apply the latest Internet technologies to bring LabVIEW to life—on the Internet or inside your organization's intranet.

LabVIEW Programming, Data Acquisition and Analysis

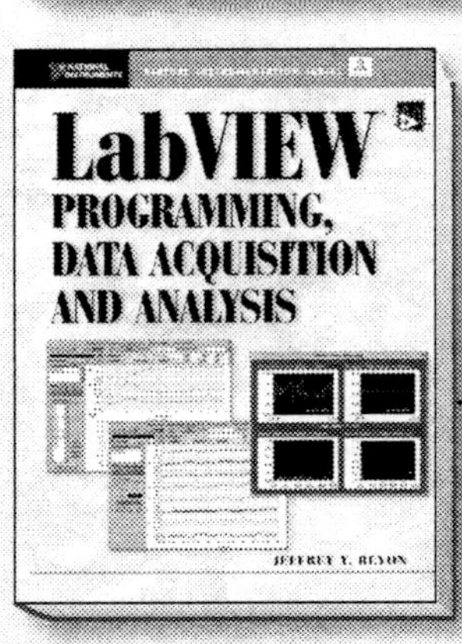

Jeffrey Y. Beyon · ©2001, Paper, 368 pp., 0-13-030367-4

For practicing engineers and students—the first book that has the structure of a college course textbook including end-of-chapter problems, with template-like examples for easy implementation.

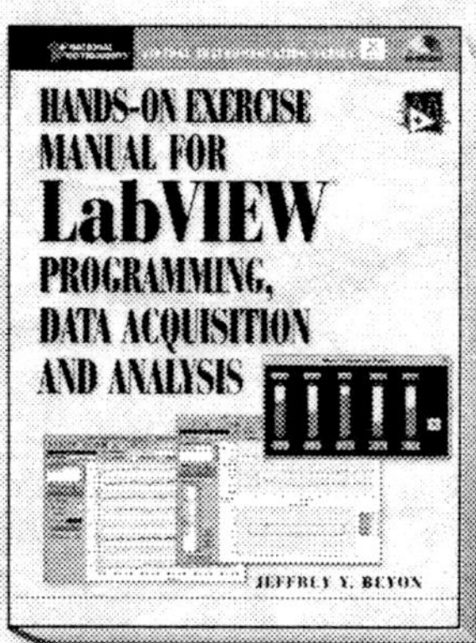

Hands-On Exercise Manual for LabVIEW Programming, Data Acquisition and Analysis

Jeffrey Y. Beyon · ©2001, Paper with CD-ROM, 100 pp., 0-13-030368-2

Structured focused practice for mastering LabVIEW programming fast!

LabWindows/CVI Programming for Beginners

Shahid F. Khalid · ©2000, Paper with CD-ROM, 656 pp., 0-13-016512-3

The first "teach yourself" guide for LabWindows/CVI!

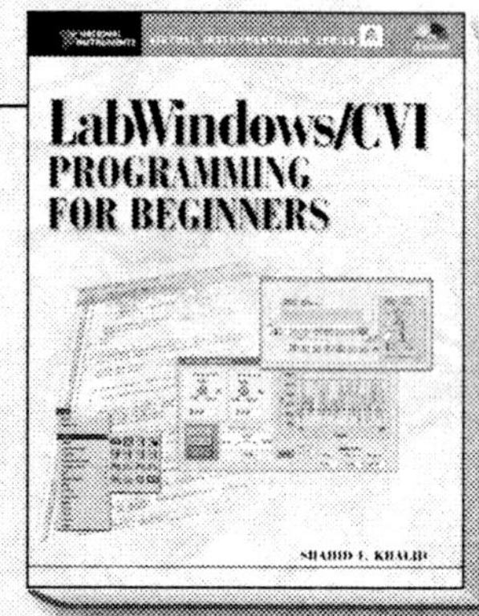

LabVIEW for Automotive, Telecommunications, Semiconductor, Biomedical, and Other Applications

Hall T. Martin / Meg L. Martin · ©2000, Paper, 272 pp., 0-13-019963-X

Practical insights, techniques and code you need to build world-class virtual instrumentation systems of your own.

Sensors, Transducers, & LabVIEW

Barry E. Paton · ©1999, Paper with CD-ROM, 350 pp., 0-13-081155-6

A guide to learning LabVIEW by hands-on experience with a wide range of applications chosen from physics, chemistry, mathematics, engineering, and medical sciences.

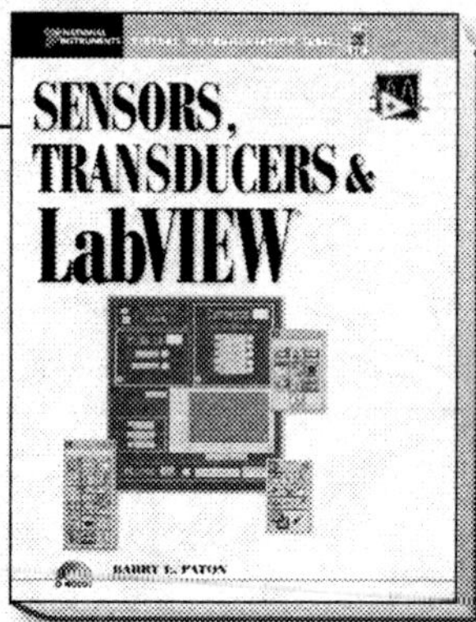

LabVIEW Signal Processing

Mahesh L. Chugani / Abhay R. Samant / Michael Cerna
©1998, Paper with CD-ROM, 688 pp., 0-13-972449-4

Understand LabVIEW's extensive analysis capabilities and learn to identify and use the best LabVIEW tool for each application—fast.

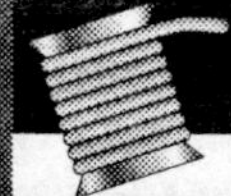

LABVIEW™ TECHNICAL RESOURCE

THE ONLY LABVIEW SUBSCRIPTION WITH VI SOFTWARE INCLUDED

ORDER FORM

TEL: 214-706-0587   FAX: 214-706-0506

CONTACT INFORMATION

Name _______________________ Company _______________________

Address ____

City _____________________ State __________ Zip / Post Code __________

Country _____________________ E-mail (required) __________________

Tel _____________________ FAX __________________

LTR PRODUCT LISTING

QTY	LTR Subscriptions and Products • Single-User License	U.S.	INTL.	EXT. PRICE
	1 year subscription (4 Issues / 4 Resource CDs)	$95	$120	
	2 year subscription (8 Issues / 8 Resource CDs)	$175	$215	
	Individual Back Issues with Resource CDs (Article Index at ltrpub.com)	$25	$30	
	VI Code Paks – *Available ONLY at ltrpub.com* (priced per pak) Download bundled paks of LTR VIs & accompanying articles by topic.	$50	$50	Order Online

QTY	LTR Library of Back Issues on CD-ROM • Single-User License	U.S.	INTL.	EXT. PRICE
	CD-ROM Library of Back Issues Ver. 4.0 (32 issues / over 350 VIs)	$375	$395	
	Upgrade CD-ROM to Ver. 4.0 (requires Version 3.0)	$99	$99	

QTY	LTR Library of Back Issues on CD-ROM • Multi-User License	U.S.	INTL.	EXT. PRICE
	Server Version CD-ROM Library of Back Issues Ver. 4.0 (5 user license)	$495	$530	

○ Please send me information on Multi-User Add-On Pak Discounts for Servers with > 6 + users.

○ Please send me information on Group Subscription discounts.

TOTAL

PAYMENT INFORMATION

✔	PAYMENT METHOD		
	VISA / MC / AMEX Card Number		Exp.
	Signature		Date
	Bill company (**U.S. ONLY**) / Fax Purchase Orders to (214) 706-0506 / P.O. Number		
	Check enclosed (**U.S. BANK ONLY** – Make check payable to LTR Publishing) (Texas residents please add 8.25% sales tax)		
	Wire / TT (**INTERNATIONAL ORDERS**) – Contact LTR for Banking Information		

ORDER ONLINE or Fax this order form to 214-706-0506. Please include a signature on all credit card orders.
Mail all check orders (US bank only) to:

LTR Publishing, Inc., 860 Avenue F, Suite 100 Plano, Texas 75074
Tel: 214-706-0587 • Fax: 214-706-0506 • email: ltr@ltrpub.com

Standard shipping is included on all orders. If you prefer, you may include your own Airborne, Federal Express, or UPS Air Collect #. Texas residents please include Texas Sales Tax at 8.25%.

LTRPUB.COM